Sand and Silicon: Science that Changed the World

Sand and Silicon: Science that Changed the World

Denis McWhan

Great Clarendon Street, Oxford OX2 6DP

Oxford University Press is a department of the University of Oxford.
It furthers the University's objective of excellence in research, scholarship,
and education by publishing worldwide in

Oxford New York

Auckland Cape Town Dar es Salaam Hong Kong Karachi
Kuala Lumpur Madrid Melbourne Mexico City Nairobi
New Delhi Shanghai Taipei Toronto

With offices in

Argentina Austria Brazil Chile Czech Republic France Greece
Guatemala Hungary Italy Japan Poland Portugal Singapore
South Korea Switzerland Thailand Turkey Ukraine Vietnam

Oxford is a registered trade mark of Oxford University Press
in the UK and in certain other countries

Published in the United States
by Oxford University Press Inc., New York

© Denis McWhan 2012

The moral rights of the author have been asserted
Database right Oxford University Press (maker)

First published 2012

All rights reserved. No part of this publication may be reproduced,
stored in a retrieval system, or transmitted, in any form or by any means,
without the prior permission in writing of Oxford University Press,
or as expressly permitted by law, or under terms agreed with the appropriate
reprographics rights organization. Enquiries concerning reproduction
outside the scope of the above should be sent to the Rights Department,
Oxford University Press, at the address above

You must not circulate this book in any other binding or cover
and you must impose the same condition on any acquirer

British Library Cataloguing in Publication Data

Data available

Library of Congress Cataloging in Publication Data
Library of Congress Control Number: 2011939856

Typeset by SPI Publishing Services, Pondicherry, India
Printed and bound by
CPI Group (UK) Ltd, Croydon, CR0 4YY

ISBN 978–0–19–964027–0

1 3 5 7 9 10 8 6 4 2

To Callie
and
Susan, Jeanette, and David

Preface

Sand is everywhere. The grains sift through our fingers while the waves break on the beach. The stress and strain of everyday life seem to drain away. But sand has a much more profound impact on our lives. We have come to depend on this ordinary material for our comfort and our commerce. In fact, we depend almost completely on it for our modern existence. Over the last century, science has taught us how to take this most common substance and create revolutionary technology. This book follows the history of these scientific discoveries and relates them to the products made from sand.

It was only 100 years ago scientists proved that materials are composed of atoms with a central nucleus surrounded by a cloud of electrons. Before that time, our knowledge of materials was based on observing how they responded to changes in temperature, pressure, electric and magnetic fields, and light. These responses defined the macroscopic properties of materials. The twentieth century brought a microscopic understanding of the atomic origin of these different properties and an understanding of the relation between the atomic structure and the properties of materials. This in turn gave scientists the ability to design materials for specific applications. First this was done on a macroscopic scale, but by the end of the century it was done on the atomic scale—literally growing materials atomic layer by atomic layer.

In the second decade of the twentieth century, scientists developed techniques to determine how the atoms in a material are arranged. In many solids, they are arrayed in a definite pattern that is repeated regularly in three dimensions, and scientists call these solids crystals. Before then, there were only speculations that the external symmetry of the minerals found in nature resulted from the microscopic symmetry that related the positions of the atoms in a crystal to one other.

Our story begins with quartz, which is a major component of sand, and with the use of quartz to detect submarines during the First World War and to make frequency stabilizers, quartz watches, sensors, and cigarette lighters. At the heart of these applications is a macroscopic property called piezoelectricity. A piezoelectric material converts mechanical oscillations to electrical oscillations and vice versa. The occurrence of piezoelectricity is related to the symmetry of a crystal, and the determination of the crystal structure of quartz led to a microscopic understanding of the origin of piezoelectricity.

Quantum mechanics was developed in the 1920s, and it explained the architecture of sand. The glue that holds materials together is the chemical bonds formed by the electrons of the different atoms in a material. Sand is the chemical compound silicon dioxide, and quantum mechanics predicts that the basic building block of sand is a grouping of atoms in which each silicon atom is surrounded by four oxygen atoms that form a tetrahedron. Sand based materials can have an infinite variety of networks of these tetrahedrons. By controlling the structure of the network, the chemical industry can make catalysts to refine gasoline, micropore filters for purifying water, and aerogel insulation for the Mars Rover.

As quantum mechanics continued to develop during the 1930s and 1940s, the electronic properties of materials were elucidated. The theory explained why some materials are metals and others are insulators. Scientists began to probe the subtle differences in the electrical conductivity of the intermediate class of materials called semiconductors, and this led to the invention of the transistor in 1946. Thus began the revolution in electronics on which our civilization has come to depend. The basis of the whole electronics industry is the conversion of sand into ultra-pure silicon. The industry then puts dopants in ultra-pure silicon in millions of microscopic regions to make computer chips. Today almost everything that we don't eat contains a silicon chip—cellphones, CD players, computers, hearing aids, TVs, and on and on. Almost every car, appliance, and toy has one or more silicon chips. The development of semiconductor physics is mirrored in the silicon devices made from sand.

Many of the discoveries of the 1960s and 1970s involved the absorption and emission of light. Light is at the core of twentieth century physics. Planck and Einstein introduced the concept that light is not continuous but is made up of quanta of energy that are called photons. The operation of solar cells, photodiodes, radiation detectors, light emitting diodes (LEDs), and semiconductor lasers involves the absorption or emission of photons. Many of these devices are made from silicon, so the physics of light is mirrored in grains of sand.

The explosion of information technology in the last three decades is based in part on optical communications. Light carries the information from computer to computer through quartz optical fibers that are made from sand. This is another example of grains of sand quite literally reflecting progress in physics at the end of the twentieth century.

When we look at a grain of sand we see how scientific discoveries have enabled us to control the structure and properties of materials and to make the products on which we depend. Behind quartz watches and catalysts is an understanding of the structure of materials derived from sand. In the end, a computer chip or a solar cell is a thin wafer of ultra-pure silicon in which an exquisite array of dopants has been implanted. To take sand and manufacture these devices from it is an outstanding achievement of twentieth-century science.

I have taken a little poetic license in the interest of pedagogy with the definition of sand. Many people associate sand with fine grains of quartz, and I have chosen to use this association as the organizing theme. (Technically sand does not refer to a particular material but to a granular material composed of grains with a certain size distribution.) Throughout the book I use the word "sand" to mean the chemical compound silicon dioxide, which is also called silica. Quartz is just one of several minerals that have the chemical composition of silicon dioxide.

Each chapter is about a different group of products that are made from sand and the twentieth-century science that led to those products. More technical aspects of the science are presented in separate boxes. A list of references is included both to the original literature and to sources and interactive graphics available on the Internet. For example, the US Patent Office has all patents on-line, and many key patent numbers are referenced. The original information is available at the Patent Office website (http://www.uspto.gov/patft/) by entering the patent number under "number search." Many of the scientific discoveries that led to products made from sand resulted in Nobel Prizes. Both the presentation address for each Nobel Prize, in which a member of the Swedish Academy of Science gives the historical significance of the prize, and the presentation given by the recipient can be found at http://nobelprize.org/nobel_prizes/physics/laureates/ by entering the year.

This book has grown out of a lifelong fascination with the new materials that are constantly being discovered by materials scientists. At the same time I have a deep seated appreciation of how these discoveries build on hundreds of years of scientific research. They didn't just happen. In an evermore complicated world, it is important to understand the scientific framework on which our technology is based so that we can make intelligent decisions. It is my hope that this brief story about sand will raise the curiosity of the reader to look beyond the sand on the beach to the science behind the technology on which we depend every day.

Much of what I have learned about materials has come from my own research done over forty years at Bell Telephone Laboratories, Incorporated (now Alcatel-Lucent USA Inc.) and at Brookhaven National Laboratory. Modern basic research is a collaborative endeavor, and I acknowledge the help of all my colleagues at Bell Labs and Brookhaven. Basic research in the physical sciences is a rewarding profession as a result of the camaraderie, the competition, and most of all the sense of wonder that comes from unraveling the secrets of nature.

Together my colleagues and I have shared that wonder, and I am indebted to them for being able to go along for the ride.

I acknowledge all my friends and colleagues who tolerated yet another discussion about sand. Each of them gave me good advice and constructive criticism. I especially thank my wife, Callie, and Leigh Sherrill and Patricia Toro who were kind enough to read each chapter time and time again as I tried to improve the book and our children Susan Tobin, Jeanette Burney, and David McWhan for encouragement and ideas. I thank my colleagues John Axe, David Litster, and David Moncton who each read the entire manuscript. A partial but not complete list of people who contributed along the way includes Dennis Burney, Stan Brown, Mike Grimes, Tom Levenson, John Parise, Jeffrey Post, Frans Spaepen, David Tobin, and Ann Woodbury. I thank the Physics Department at MIT for giving me a guest appointment during the time I did the research for the book. Finally, I thank the staff at Oxford University Press including the Physics Editor Sonke Adlung, the Production Editor Victoria Mortimer, and Subramaniam Vengatakrishnan at SPi Global for making this book a reality.

Table of Contents

List of Figures   xiii

1. Submarines, Clocks, and Sensors   1
2. The Architecture of Sand   21
3. How Pure is Pure?   47
4. Impurities are Key   65
5. The Sun Shines Bright   87
6. How Small is Small?   101
7. Through the Looking Glass   117
8. Sand is Everywhere   129

Bibliography   133
Index   139

List of Figures

1.1	Symmetry of quartz	2
1.2	Piezoelectricity	4
1.3	Curie apparatus to determine radioactivity	6
1.4	Submarine detection	9
1.5	Synthetic quartz crystal	12
1.6	Seiko Pulsar and Casio Data Bank quartz watches	13
1.7	Structure of α and β phases of quartz	16
1.8	Unit cells on a cubic lattice	19
1.9	Symmetry of oxygen atoms in β-quartz	20
2.1	Bragg's Law for diffraction of x-rays	28
2.2	Structure of silicon	29
2.3	X-ray powder diffraction patterns (bar codes) of silicon and quartz	30
2.4	Materials derived from sand	33
2.5	Directional bonds in silicon	34
2.6	The structure of glass and tridymite	35
2.7	CORELLE® glass ceramic dinnerware	37
2.8	Tunnels in the zeolite, Faujasite and in Tridymite	39
2.9	Insulating properties of aerogel	43
2.10	Electron micrograph of opal	44
2.11	Silica skeletons of diatoms	45
3.1	Purification of silicon by zone refining process	49
3.2	Purification of silicon by Siemens process	50
3.3	Czochralski and float zone growth of single crystals of Si	51
3.4	Single crystal boule of silicon	52
3.5	Transition from the liquid to the vapor phase	53
3.6	Vapor pressure of acetone and water	56
3.7	Phase diagram for solution of acetone and water	59
3.8	Pressure versus volume for a liquid that is boiling	63
4.1	Quartz-based gemstones	66
4.2	The discovery of the p–n junction	69
4.3	p- and n-doping of a semiconductor and the p–n junction	71
4.4	Early junction transistor	77
4.5	Comparison of junction and field effect transistors	78

4.6	Comparison of 1971 and 2000 INTEL® microprocessors	80
4.7	Comparison of experiment to establish the nucleus and ion implantion	85
5.1	Solar cell powered Telstar I communications satellite	88
5.2	8 MW solar power system by SunEdison LLC in Alamosa, Colorado	89
5.3	Comparison of Rayleigh Jeans and Planck's Laws with the solar spectrum	92
5.4	High efficiency solar cell	95
5.5	Triple junction solar cell	96
5.6	p–i–n photodiode to detect light in optical communications	98
5.7	CCD detector	99
5.8	STAR silicon detector at RHIC	99
6.1	Strain modulated heterojunction bipolar transistor and field effect transistor	104
6.2	Heteroepitaxial growth of one material on top of another material	109
6.3	Multilayer with alternating regions of GaAs and AlAs	109
6.4	Transmission electron microscope image of a multilayer crystal	110
6.5	p-AlGaAs/GaAs/n-AlGaAs laser diode	115
7.1	Quartz optical fibers	118
7.2	Guiding of light in a quartz optical fiber	122
7.3	Drawing an optical fiber	124
7.4	Production of quartz preforms used to draw optical fibers	126

1
Submarines, Clocks, and Sensors

Sand means different things to different people. To the child on the beach or in a sandbox, it is a source of endless pleasure. It is perfect for loading and hauling and building sandcastles. To the vacationer the beach sand is for running, walking, or just soaking up the sun. To the golfer it is the sand trap where the ball lands after an errant stroke.

To the scientist and the engineer sand provides a different type of sandbox. Sand is the starting material for many useful products. There are the obvious things like glass for containers and windows and concrete for the construction of buildings and roads. But it is with the discoveries of twentieth century science that the sophistication of the products made from sand began to evolve at an accelerating rate. As sand is so abundant, it is an ideal substance—an attractive sandbox—to illustrate how these products have developed from the discoveries of twentieth century science.

Sand is defined as gritty particles of worn or disintegrated rock with grains varying in size from tenths of a millimeter to several millimeters. A major component of sand is the mineral quartz. Sand often contains other materials so the composition and color of sand varies depending on its origin (Greenberg 2008). There are the black sands of Hawaii and the white sands of Cape Cod. In addition to quartz these sands contain the remnants of volcanic eruptions and ground up seashells respectively. The sand in the National Sand Dune Monument in Colorado contains ground up rocks that are composed of various silicates in addition to quartz. Geologists study grains of sand to understand the processes that control the shape of the Earth's surface and to reconstruct the conditions under which sandstone was formed (Siever 1988 and Welland 2009).

The crust of the Earth is approximately 25% silicon and 50% oxygen by weight so it is not surprising that quartz, which is composed of one silicon atom for every two oxygen atoms, is one of the most common components of sand. There are a number of minerals with the composition silicon dioxide that have the generic name silica, and some of these will be discussed in later chapters. Throughout this book I use the term "sand" loosely to refer to all forms of silica but particularly to quartz.

Every natural history museum and rock shop has a collection of crystals of quartz. The centerpiece of the collection of the Smithsonian National Museum of Natural History in Washington D.C. is the world's largest perfect sphere of quartz (the Warren Sphere) that weighs 107 lbs (49 kg) and is 12.9 in (32.7 cm) in diameter. There is a display of quartz crystals arrayed in concentric rings at the American Museum of Natural History in New York. The colors evolve from a clear rock crystal at the center to a seemingly endless variation in the color of different quartz minerals like amethyst and citrine. (The colors of quartz gemstones are discussed in Chapter 4.)

Crystals of quartz are admired for their beauty, but they are also very important to industry. Sand in the form of little slices of quartz is in cell phones, computers, various sensors, quartz watches, and cigarette lighters. It is estimated that 10 billion quartz crystal devices are produced and installed in equipment every year (US Geologic Survey 2008).

These devices use a property of quartz called piezoelectricity, a Greek word piezein meaning to press and the word electricity. In other words, it is electricity resulting from pressure. A transducer uses the piezoelectric effect to convert mechanical oscillations into electrical oscillations or vice versa.

The occurrence of piezoelectricity depends on the symmetry of the crystal. One of the themes in physics in the twentieth century is the importance of symmetry and how the properties of crystals are affected by symmetry. In Fig. 1.1 a crystal of quartz is compared to a textbook drawing that illustrates the symmetry of quartz. If the crystal is rotated by 120° and 240° around the z axis or by 180° around the x axis, it will look identical to the unrotated crystal. Only certain combinations of symmetry elements allow crystals to exhibit piezoelectricity.

 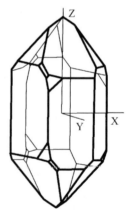

Figure 1.1 Quartz crystal and an idealized drawing showing the symmetry of quartz. The faces of the crystal are related to one another by a three-fold and a two-fold axis of rotation along the z and x axis respectively. Quartz can be piezoelectric because it does not have a center of symmetry. Each face of the crystal does not have an equivalent face on the opposite side of the crystal.

Let's follow the use of quartz crystals in the twentieth century and then look at the origin of the piezoelectricity in quartz. Many of us have heard the ping of the sonar in "The Hunt for Red October" and have seen ultrasound images of the fetus in the womb. These technologies grew out of submarine detection programs in the First World War that were based on quartz transducers. In the 1920s, quartz found applications—based on piezoelectricity—as frequency standards and filters to separate specific frequencies. These applications were important to the nascent telephone and radio communications industries and are now integral parts of the modern telecommunications and electronics industries. The frequency standards led to the development of quartz clocks and then quartz watches and different types of sensors.

Piezoelectricity was discovered by the brothers Pierre and Jacques Curie in 1880 (Curie and Curie 1880). The name "Curie" is well known for all the important scientific discoveries that different members of the family made in the late nineteenth and early twentieth century. Between Pierre, his wife Marie, and their daughter Irène, Curies won or shared in four Nobel prizes.

Pierre Curie was born in Paris on 15 May 1859. He graduated in only two years from the Sorbonne with a degree in physics at age eighteen. Rather than pursuing an advanced degree immediately, he took a job as a laboratory assistant to help support the family (Brian 2005).

Pierre and Jacques both worked for Charles Friedel in the laboratory of mineralogy at the Sorbonne. At the time, Friedel was interested in an unusual property of some crystals where a weak electric voltage develops across the crystal when it is heated. This phenomenon is now known as pyroelectricity from the Greek for fire and electricity. Friedel asked the Curie brothers to carry out a systematic study of the occurrence of pyroelectricity. They found that only certain types of crystals were pyroelectric, and they related the occurrence of pyroelectricity to crystals with specific types of symmetry (as discussed below a crystal belongs to one of 32 macroscopic symmetry classes or point groups and only 10 of these allow pyroelectricity).

The brothers wondered whether, if there were a relation between electric charge and temperature, then would there also be a relation between electric charge and pressure? They tested many different crystals such as quartz and found that some of them became electrically polarized with the application of pressure. Their early experiments were quite primitive. They would cut a plate out of a crystal with a jeweler's saw and then squeeze the plate between two pieces of tinfoil in a vise to see if an electric polarization developed (Fig. 1.2). The brothers had discovered a new, previously unknown property, and they called it piezoelectricity by analogy with pyroelectricity.

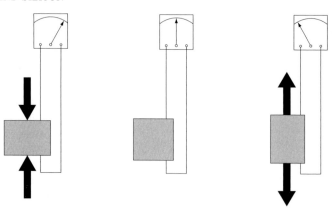

Figure 1.2 A piezoelectric crystal becomes electrically polarized when it is compressed. The voltmeter connected to the faces of the crystal shows a positive or negative voltage when the crystal is compressed or stretched. The piezoelectric response depends on the specific symmetry of the crystal.

Another professor at the Sorbonne, Gabriel Lippman, predicted that the reverse effect should also exist. If an electric voltage is applied to a piezoelectric crystal, then the crystal should expand and contract depending on the polarity of the voltage. The Curie brothers demonstrated this effect the following year.

To study pyroelectricity and piezoelectricity, Pierre Curie had to measure very small electric charges. The traditional method to measure charge is with an electrometer. A common demonstration in physics classes is the generation of electrical charge by rubbing a cloth over an ebony rod and then touching the rod to a gold leaf electrometer. This consists of a pair of gold leaves that are suspended from a central metal contact. The gold leaves bend away from each other as charge is transferred from the rod to the contact. The charge flows to both leaves and like charges repel each other. The separation of the gold leaves is a measure of the amount of charge.

Pierre Curie developed a much more sensitive electrometer. It relied on what is known as a torsional balance. A metal disk is cut into two opposing quarter disks that are suspended from a quartz fiber. The quarter disks are placed between the plates of four capacitors that are arranged in a circle. The difference between the unknown charge on two opposing capacitors and a known charge on the other two capacitors results in a torque that causes the disk to rotate. In turn this twists the quartz fiber. A beam of light is reflected from a mirror attached to the fiber, and the reflected light moves as the quartz fiber is twisted.

Extremely small changes in charge can be detected by the shift in the reflected light (Barbo *et al.* 2004).

Over the next few years, Pierre Curie systematically studied a large number of crystals with different types of symmetry. He derived and demonstrated the relation between the symmetry of a crystal and its piezoelectric response. In 1882 Pierre moved from the Sorbonne to become a lecturer at the Ecole Supérieure de Physique et Chimie Industrielles de Paris (ESCPI). He began studying the magnetic properties of materials as a function of temperature for his thesis. He discovered that when magnetic materials are heated, there is a critical temperature above which they cease to be magnetic. This temperature is now known as the Curie temperature.

After three years of research, Pierre wrote his thesis and was awarded a Doctor's degree in March 1885. During that time, he had been teaching at the school and was particularly attracted to a young Polish girl, Marie Sklodowska, who was one of his students. Once he had received his degree and had a steady income, Pierre felt that he could support a family and asked her to marry him. They were married on 26 July 1885. This was the beginning of one of the most famous and successful husband and wife collaborations in the history of science.

For Pierre and Marie, the timing could not have been more favorable. One year later, in 1896, Henri Becquerel discovered a new form of radiation emanating from uranium compounds. Here were materials that spontaneously produced some new and strange form of radiation. Nothing like this had ever been seen before. Science in the nineteenth century was built on the principle of the conservation of energy. There had to be a balance of energy, and to emit light or electrons or x-rays from a material, energy had to be supplied to the material. The uranium compounds emitted radiation without an obvious source of energy. What was the origin of the radiation? Here was an opportunity for Marie to do research at the forefront of science for her thesis.

Becquerel was studying the origin of the phosphorescence that is observed in some uranium materials when by chance he observed what became known at Becquerel rays. A phosphorescent material absorbs light and then slowly re-emits the light over a period of time. Becquerel had the idea that phosphorescence also resulted in the emission of x-rays. To study this phenomenon, Becquerel exposed a uranium compound to sunlight, and then placed it on top of a photographic plate that was shielded from light by black paper. After a period of time he developed the plate and found that the plate showed an outline of the uranium sample. He concluded incorrectly that since the black paper blocked the light, x-rays emitted by the uranium must have fogged the photographic plate.

In one of those accidents of history, Becquerel was repeating the experiment, but the sun was not shining. He developed the plate anyway and found the same fogging of the plate. As the sample had not been exposed to light to stimulate the phosphorescence, some new form of radiation had to have emanated from the uranium compound itself in order to darken the photographic plate. Becquerel then found that the radiation ionized the air causing it to conduct electricity. This new radiation was initially called Becquerel rays, but it was later discovered that in fact Becquerel rays were composed of three different types of radiation that are now called alpha, beta, and gamma radiation. The latter two are electrons and high energy x-rays respectively (Box 4.2).

Is uranium unique or are there other materials that might emit Becquerel rays? The Curies decided to search for other materials. To do this they needed a method to rapidly test many different materials, and quartz played an important role in the testing. They decided to use the conductivity of the ionized air as a measure of the presence of this new type of radiation. They could test for the presence of the radiation by the presence or absence of the electric current. To do this they needed an apparatus that was sensitive to the extremely weak current produced by the radiation. Pierre had already developed the necessary parts. He had developed a sensitive electrometer, and he could produce a known quantity of electric charge in the form of the piezoelectric response of a quartz crystal. The Curies' idea was to use the electrometer to compare the charge developed by the ionized air with the charge produced by the piezoelectric crystal.

The apparatus developed to screen materials for spontaneous radiation is sketched in Fig. 1.3. A weight pulls on the quartz crystal thereby producing a known amount of electric charge as indicated by the electrometer. The substance to be tested is placed between the plates of a capacitor. The radiation ionizes the air producing charged particles that result in a very weak current flowing between the plates of the capacitor.

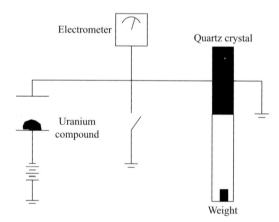

Figure 1.3 Marie and Pierre Curie's apparatus to detect radioactivity. A known charge is produced by stretching a piezoelectric quartz crystal. Radioactive decay ionizes the air in the capacitor producing a charge that increases with time. The strength of the radioactivity is the time needed to compensate for the piezoelectric charge when the weight is removed. (Adapted from Barbo et al. 2004.)

In order to determine the strength of the radioactivity in the material, the rate at which the charge flows across the capacitor has to be measured. Initially the switch is closed so the current passes through the capacitor and returns to ground. When the switch is opened, the electrometer registers the charge produced by the piezoelectric response of quartz to being pulled by the weight. In addition, charge is produced by the ionization of the air in the capacitor. Marie would remove the weight from the quartz, and the electrometer would show the loss of charge as the piezoelectric voltage went to zero. With a stopwatch, she would measure the time it took for the radioactivity to produce enough charge to compensate for the charge originally developed by the piezoelectricity of quartz. The strength of the radioactivity is the initial charge produced by the quartz divided by the time it took to balance the electrometer.

In 2006, which was the 100th anniversary of the death of Pierre Curie, scientists at ESCPI reassembled the apparatus that had been built for the Curies by the instrument maker Charles Beaudouin and recreated the original experiment (Barbo 2004). They found that they were able to measure a current of less than 10^{-13} amperes. Modern electrometers based on solid state electronics are only 100 times better (10^{-15} amperes) than the apparatus used a century ago by the Curies.

The Curies used their apparatus to test many different minerals for radioactivity and found that only those containing the elements uranium and thorium showed radioactivity. Moreover, the strength of the radioactivity was proportional to the amount of those elements in the compounds, indicating that the Becquerel rays had to come from those specific elements.

However, pitchblende, a mineral that contained uranium and thorium, showed erratic behavior. There was much more radioactivity than expected from the amount of uranium and thorium, and the amount of radioactivity varied from sample to sample. This suggested that there was something else that was producing Becquerel rays in addition to the uranium and thorium. Marie speculated that there must be small amounts of impurities in the pitchblende and that they were new previously unknown elements.

The challenge was to identify the extra radioactive component in the pitchblende. Over many years, chemists developed a sequence of chemical reactions to separate compounds of different elements from each other. The sequence is based on differences in the solubility of chemical compounds so, in a reaction, ions of elements in one column of the periodic table remain in solution while those of another column in the table separate or precipitate out of the solution. As a chemistry major in college, I remember the course in chemical qualitative analysis and the endless hours it took to methodically go through the recipe, reaction by reaction, to analyze an unknown material.

The Curies had to carry out a similar sequence of chemical separations. At each step along the way they determined whether the radioactive fraction was in solution or in the precipitate. This allowed them to follow the radioactivity as they worked through the chemical analysis. They found two different components with enhanced radioactivity. One had chemical properties similar to bismuth and the other to barium. These two elements are in different columns of the periodic table from thorium and uranium, so the enhanced radioactivity did not emanate from either of them. As Marie had guessed initially, there were impurities in pitchblende that were new elements. Marie and Pierre gave them the names polonium and radium respectively.

When one looks back at these experiments, they are truly remarkable. The Curies separated minute quantities of these new elements from the pitchblende. Furthermore, they did not have today's sophisticated analytic equipment nor did they even have an adequate laboratory. Their work was done in a shed behind the main building of the Institute.

The scientific community was skeptical of their results. In the end seeing is believing, so the Curie's had to isolate enough of each of the elements to be able to show their colleagues a macroscopic sample. This required processing literally tons of pitchblende to separate weighable amounts, but with Marie's perseverance, they proved the existence of polonium and radium. They determined the atomic weight of each element to show that it was different from any other element in addition to having different chemical properties. Marie and Pierre Curie shared the Nobel Prize in Physics in 1903 with Henri Becquerel for their studies of radioactivity.

Unfortunately for the world of physics, Pierre met an untimely death in 1906 at the age of 47 when he was run over by a horse drawn carriage. His legacy continued not only through his wife and daughter, but also through his student Paul Langevin who used the piezoelectricity of quartz in his apparatus for submarine detection.

The first practical use of the piezoelectricity of quartz arose in the First World War. England, France, and the United States all started research programs to detect submarines because they were having a devastating effect on shipping. Initially the English and the Americans concentrated on improving hydrophones, which are underwater microphones used to listen for sounds under water. They had been developed as navigational aids in the late 1800s. Underwater bells were placed under lightships or near lighthouses and other navigational obstacles. The sound of the bell was received by a hydrophone mounted under the bow of the ship thereby warning it of a nearby hazard. In principle it should be possible to listen for sounds emanating from submarines, but it proved to be difficult at the time to develop a reliable system.

In 1915 the French scientist, Paul Langevin, tried a different approach. Langevin, like his mentor Pierre Curie, had begun his career studying the magnetic properties of materials. He was quite familiar with the work of Pierre Curie on piezoelectricity and decided to use the piezoelectricity of quartz to detect submarines. Instead of a passive system that listened for the sound emanating from the submarine, he and Constantin Chilowsky devised an active system. A transmitter sends out a sound wave that is reflected off an object, and the reflected wave or echo is detected by a receiver. Sound with frequencies higher than those that can be detected by the human ear, ultrasonic waves, propagate long distances in water. To generate ultrasonic waves, Langevin used the piezoelectric response of a quartz crystal to convert a high frequency electric wave into a high frequency mechanical oscillation. In turn this caused a metal plate to vibrate, thereby generating an ultrasonic wave (Fig. 1.4).

Langevin describes the invention in his patent application:

> "In Patent No. 1,471,547, granted in the names of Chilowsky and the applicant, an apparatus has been described by which the directive emission and reception of elastic waves of high frequency in water are obtained by means of electric oscillations and applications of these means for secret and directed submarine signaling, for detecting submarines and submarine mines and for protecting vessels against reefs, sandbanks, icebergs and collisions of any kind by utilizing the echo produced or the shadow thrown by the obstacle to be detected.... The present invention consists in means which attain the same object by utilizing the piezoelectric properties of quartz in order to obtain the transformation of electric oscillations of given frequency into elastic waves of the same frequency and vice-versa." (Langevin 1920).

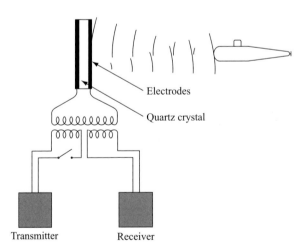

Figure 1.4 The invention of ultrasonic imaging by Paul Langevin. The transmitter induces a pulse of alternating electric voltage across the electrodes causing the piezoelectric quartz crystal to produce a mechanical oscillation of the same frequency. The resulting ultrasonic wave propagates in the water and is reflected off the submarine. The reflected wave in turn causes the quartz crystal to produce an alternating electric voltage that is detected by the receiver. (Adapted from Langevin 1920.)

Langevin slowly improved his device and, after the First World War, started a company to manufacture depth finders. R. W. Boyle and his group at the British Board of Invention and Research built on Langevin's work to make the first practical submarine detection system using quartz transducers in 1918 (Hackmann 1986).

Over the ensuing century ultrasonic imaging has become ubiquitous, but it all started with the early experiments of Langevin using quartz. In fact on the rue des Ecoles in Paris, there is a small park named "Square Paul Langevin" that has a plaque commemorating his achievement.

A second application of the piezoelectricity of quartz, dating from the early 1920s and continuing to this day, is in radio and telephone communications. A radio station transmits a signal at a specified frequency. A land line telephone call is transmitted through copper wires and a cell phone call is transmitted through the air by modulating a carrier wave that has a specified frequency. In each case the frequency of the respective transmitters have to be stable and not drift with time. Walter G. Cady at Wesleyan University (Cady 1920) and Alexander McLean Nicolson at the research department of the Western Electric Company (Nicolson 1918) incorporated piezoelectric crystals into resonant electrical circuits to provide stable frequencies. The economic value of this invention to the budding communications industry led to a protracted patent fight. Walter Cady lost the battle to Alexander Nicolson and the Western Electric Company.

In a resonance the amplitude of a signal is greatly enhanced at a specific frequency. The notes on musical instruments like pipe organs, horns, and flutes are resonances that are controlled by the length of the air column. The air column acts as a resonator to reinforce and control the vibration at the frequency of the note. Similarly a crystal or a tuning fork has resonant frequencies that depend on the dimensions of the crystal and on its intrinsic properties. In a quartz oscillator, the piezoelectric crystal produces a voltage that oscillates at the resonant frequency of the crystal. The piezoelectric oscillator locks the frequency of the oscillator to the natural resonant frequency of the quartz crystal and is an accurate frequency standard (Heising 1946).

A crystal resonator can also be used as a filter to pass only a desired band of frequencies. In telephone communications or AM radio, many conversations or radio stations, each with their own assigned band of frequencies, are transmitted down copper wires or over the airwaves. A crystal filter allows a specific band of frequencies to be detected by the receiver while suppressing all other nearby frequencies.

Quartz is the material of choice for crystal oscillators and filters because the resonance is particularly narrow. Since the 1920s quartz oscillators and filters have been standard components of telephone and radio communications equipment. Today quartz resonators are found in cell phones, TVs, and other consumer electronics that need to operate at stable frequencies.

The production of quartz resonators really expanded as the Second World War approached. The military needed large numbers of transmitters and receivers to communicate between planes, ships, and ground bases, and each of them required a quartz resonator. During the first half of the twentieth century, naturally occurring quartz crystals were used by industry to make resonators, and these were mined principally in Brazil. With the increase in demand for quartz resonators, attempts were made to grow synthetic (or cultured) quartz crystals on an industrial scale.

Because of the extensive use of quartz in the Bell System, a program was initiated at Bell Labs., the research arm of the Bell System, to grow quartz crystals. A. C. Walker and Ernie Buehler developed a hydrothermal process to grow crystals of suitable quality for piezoelectric devices (Walker and Buehler 1950). As the name implies (hydro = water and thermal = heat) hydrothermal synthesis involves heating a mixture of silica sand or quartz microcrystals, water, and sodium hydroxide in a high pressure vessel called an autoclave to between 375 ° and 450 °C. Under these conditions the pressure inside the autoclave is several thousand atmospheres.

An autoclave is built to withstand a certain maximum pressure, but above that pressure the vessel might fail resulting in an explosion. Just as hot water heaters have relief valves, the autoclave has a rupture disk as a precaution. It is designed to burst at a lower pressure. However, these pressures are substantially higher than those in a hot water heater, and a stream of supercritical water blowing out of a ruptured disk is a safety concern. Partly because of this, the laboratory that Walker and Buehler used was located in a separate building behind one of the main buildings of the Murray Hill, New Jersey research complex. There is a story that the Harvard Professor, Percy Bridgman, (who won the Nobel Prize for his studies of the properties of materials at high pressure), would always take a hop when he walked past the area in his laboratory that was in line with the seal in his high pressure apparatus.

Walker and Buehler designed a thin walled cell to contain the corrosive liquid that fitted snuggly into the autoclave. Sand is placed at the bottom of the cell and seed crystals are suspended from a rack at the top of the cell. The autoclave is placed inside a furnace and heated so that there is a temperature gradient with the top being slightly colder than the bottom. The sand slowly dissolves and is transported to the seed crystals by convection. The crystals grow slowly over a period of weeks, and a typical quartz crystal grown by this process is shown in Fig. 1.5.

Today, quartz crystals are grown using similar techniques by many companies around the world to meet the demands of industry. The US Geologic Survey of commodities refers to these crystals as "cultured quartz crystals." Although natural quartz crystals were used in the first half of the twentieth century for electronic applications, cultured crystals were used almost exclusively by the end of the century (http://minerals.usgs.gov/minerals/pubs/commodity/silica/mcs-2010-quart.pdf).

Figure 1.5 Synthetic (cultured) quartz crystal grown at Bell Telephone Laboratories. In the hydrothermal growth of quartz, sand dissolves in water at high temperature and high pressure and is transported by convection to seed crystals that grow into high quality crystals for use in piezoelectric devices.

Quartz clocks and watches are another application of crystal resonators. Walter Marrison at Bell Labs is credited with the invention of the quartz clock in 1927 (Marrison 1930). Prior to this invention, precise knowledge of the time was based on exquisite mechanical clocks. Dava Sobel in her book *Longitude* describes the struggle to develop accurate chronometers for navigation in the eighteenth century (Sobel 1995). The chronometer developed by John Harrison in 1761, which she relates, represented a remarkable achievement. It had an accuracy of about a fifth of a second per day. Further improvements led to pendulum clocks such as the Shortt clock developed in 1921 that had an accuracy of hundredths of a second per day.

Marrison realized that a quartz resonator provided a stable frequency, and therefore a clock could be built by reducing the frequency and using the low frequency signal to drive a synchronous motor to turn the clock mechanism.

The resonant frequency of a quartz crystal is an intrinsic property of the crystal, and the accuracy of a quartz clock is limited by the variation of the resonant frequency with temperature. The orientation of the quartz crystal plate is chosen to minimize that dependence, and the clock is maintained at a constant temperature. Under these conditions quartz clocks can achieve relative accuracies of fractions of a second per month.

For several decades quartz clocks were used as time standards, but in the 1950s even more accurate atomic clocks were developed. They depend on an intrinsic property of an atom rather than on the vibration of a crystal, so they are inherently more stable. The cesium atomic clock maintained by the National Institute of Standards and Technology has an accuracy of 30 billionths of a second per year (see: A Walk Through Time at http://www.nist.gov/pml/general/index.cfm).

The development of the quartz watch came after the Second World War with the invention of the transistor and the integrated circuit. The electronics of the quartz clock could be miniaturized, and in the late 1960s both Seiko in Japan and CEH (Centre Éléctronique Horogère) in Switzerland introduced the first quartz wrist watches (Fig. 1.6). With the development of liquid crystal displays, all moving mechanical parts were eliminated and the digital quartz watch became another everyday item (Fig. 1.6). Fine mechanical watches are luxury items while quartz watches with accuracies of half a second per day are purchased for tens of dollars at Walmart.

Quartz resonators are at the heart of many types of sensors. Instead of using the resonator to provide a frequency standard, a quartz sensor measures changes in the resonant frequency that result from changes in the mass or temperature of a quartz crystal. For example, a quartz crystal microbalance (QCM) is able to detect the addition of a single monolayer of atoms onto the surface of a crystal of quartz. This is a change in mass of only 1×10^{-8} grams/cm^2.

Figure 1.6 Casio "Data Bank" and Seiko "Pulsar" quartz watches. The piezoelectric response of a quartz crystal stabilizes the frequency of the oscillator circuit that controls the watch.

The principle behind a QCM is analogous to plucking the string of a violin. The string oscillates back and forth in a direction that is perpendicular to the axis of the string. This is referred to as a transverse wave. The frequency of vibration of the string depends on its length, the tension on the string and on the mass of the string per unit length. In a quartz plate, a transverse wave is produced by applying a stress parallel to the surface of the crystal producing a wave that oscillates in a direction perpendicular to the surface of the crystal. The resonant frequency of the wave is sensitive to the small change in the thickness of the quartz plate when a thin film is deposited on the surface (Ward and Buttry 1990).

The change in the resonant frequency of a quartz plate with changes in temperature can be used to make thermometers. The properties of quartz are anisotropic, that is, they are different for different directions in the crystal. Devices are made by cutting plates out of a crystal. The orientation of the plate with respect to the natural faces of the starting crystal is referred to as a "cut." Different cuts are used if the desired piezoelectric response is the result of a simple compression or the result of shearing the plate. For use as a frequency standard, orientations are chosen in which the resonant frequency is almost independent of temperature. However, there are cuts where the resonant frequency varies linearly with temperature, and the piezoelectric voltage is a measure of the temperature. Thermometers with accuracies of millidegrees and even microdegrees use quartz oscillators as the sensing element (Martin 1976).

Quartz is used to make accelerometers and pressure sensors (Errol *et al.* 1988). These applications are based on the change in the piezoelectric voltage that results from mechanical deformation. When an object is accelerated, it produces a force, and according to Newton's law, force is equal to the mass multiplied by the acceleration. If the object experiencing acceleration presses against a quartz crystal, then the resulting force compresses the quartz crystal giving rise to a piezoelectric voltage. This voltage is directly proportional to the acceleration. Accelerometers find wide application in industry as vibration sensors in large structures like bridges, automobiles, and aircraft.

Another type of accelerometer is based on silicon made from sand. Miniature cantilever beams made out of silicon bend when an attached seismic mass is accelerated. These devices are called micro-electromechanical systems, MEMS, and they are made using silicon technology similar to that discussed in Chapter 4. MEMS devices are replacing quartz accelerometers and are used, for example, in cell phones to sense the orientation of the phone.

Single crystals of quartz are used in many different piezoelectric devices, but what leads to piezoelectricity? Why is quartz piezoelectric while many other minerals aren't? In most materials there is a balance between the positive charge of the nuclei and the negative charge of the electrons in the atoms that make up the material. In order for a quartz plate to spontaneously develop an electric polarization as a result of being compressed, this balance must be broken. The net positive charge must shift relative to the net negative charge in the material. For this to happen a piezoelectric material must be inherently different from a non-piezoelectric material, and that difference is related to the symmetry of the material.

Crystals are classified by the symmetry that they exhibit (Box 1.1). The faces of a crystal are related to each other by different types of symmetry operations, and the symmetry operation that is important for piezoelectricity is a center of symmetry. If every face on a crystal has an equivalent face on the other side of the crystal, then the crystal is said to have a center of symmetry. If opposite faces are equivalent, then the material can't be piezoelectric because there would be no reason for a charge imbalance to occur as a result of compression. Only crystals that do not have a center of symmetry can be piezoelectric.

In the classification of crystals by their external symmetry, there are 32 unique combinations of symmetry elements, and only 20 of these combinations allow piezoelectricity. For example, minerals composed of silicon dioxide adopt different crystal structures depending on the temperature and pressure at which the mineral was formed. Some of these minerals (tridymite, coesite, and stishovite) have centers of symmetry, and some (quartz and cristobalite) do not. Only quartz and cristobalite can exhibit piezoelectricity.

In order to understand the atomic origin of piezoelectricity, a microscopic classification of symmetry is needed. In this classification, the smallest unique grouping of atoms that are contained in a volume in space called the unit cell is defined. The atoms in the unit cell are related to one another by the macroscopic symmetry operations plus additional symmetry operations that combine these operations with translations of the atoms inside the unit cell. There are 230 unique combinations of these symmetry operations, and they are referred to as space groups. The macroscopic classification of crystals is into 32 unique classes, and the microscopic classification of crystals is into 230 unique space groups.

As will be described in Chapter 2, techniques were developed at the beginning of the twentieth century to look inside a grain of sand (quartz) and to determine the positions of the atoms in the unit cell. In turn, this led to an understanding of the origin of piezoelectricity at the atomic level. To illustrate the atomic model of piezoelectricity, the positions of the atoms in the unit cell of quartz that occurs at high temperature (β-quartz) are compared with the positions of the atoms at low temperature (α-quartz).

In 1926 (about the same time that sonar and quartz clocks were being developed) Lawrence Bragg and R. E. Gibbs determined the structures of both α-quartz and β-quartz (Bragg and Gibbs 1926 and Gibbs 1927). The crystal structure of β-quartz is shown at the top right of Fig. 1.7. Each of the silicon atoms (large spheres) is bonded to four oxygen atoms (small spheres). The oxygen atoms can be thought of as being at the corners of a tetrahedron. Each oxygen atom is shared between two

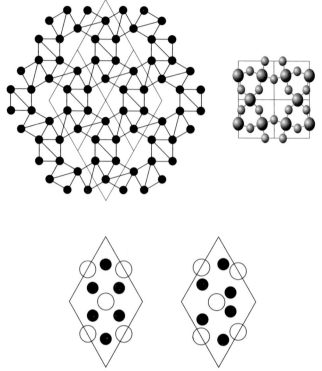

Figure 1.7 In the structure of β-quartz (top right), each silicon atom (large sphere) is bonded to four oxygen atoms (small spheres) that are at the corners of a tetrahedron. The tetrahedrons share corners to form a hexagon, and the structure can be viewed as a network of connected tetrahedrons (top left). The smallest grouping of atoms, the unit cell, that define the structure are outlined. The atomic origin of piezoelectricity is illustrated by comparing the unit cells of β-quartz (left) and α-quartz (right). In α-quartz the lower symmetry allows a charge imbalance to develop between the positively charged silicon atoms and the negative charged oxygen atoms when the unit cell is compressed in the horizontal direction. In β-quartz the higher symmetry prevents a charge imbalance from occurring.

tetrahedrons. When the three dimensional structure is projected onto the plane of the paper, adjacent tetrahedrons appear as a hexagon with the edge of each tetrahedron forming a side of the hexagon.

The structure of β-quartz can be visualized as a network of tetrahedrons, as illustrated in the top left of Fig. 1.7. (As discussed in Chapter 2, the architecture of sand is based on networks of tetrahedrons.) Superimposed on the interconnected hexagons is the outline of the unit cells that define the structure of quartz in the plane of the paper. The unit cell is defined by two axes of equal length that are 120° apart. The third axis has a different length and is perpendicular to the paper. The structure of β-quartz is completely described by the positions of the atoms in one of the unit cells that are outlined in the figure. The macroscopic structure is built up by repeating the unit cell, much like tiling a floor but in three dimensions. Four adjacent unit cells are outlined in the figure.

The origin of the piezoelectricity in quartz is the shift in the distribution of electric charge within the crystal when it is compressed. It is extremely difficult to reliably calculate this shift even with present day computing power, but a qualitative picture can be developed from the early models of the chemical bonds that hold the atoms in a crystal together. They are based on two extremes. In the model proposed by G. N. Lewis in 1923, the bonding electrons in silicon and oxygen are shared to form what are called covalent bonds. In the model proposed by Linus Pauling in 1929, the bonding electrons in silicon are transferred to the oxygen atoms so that the silicon has a charge of plus four and the oxygen atoms have a charge of minus two. In this case the atoms in the crystal are held together by the ionic attraction of the positively and negatively charged atoms.

There is controversy to this day in the scientific literature about the true nature of the silicon–oxygen bond. Both models are overly simplified pictures of the chemical bonding in silicon dioxide, and it is reasonable to assume that the true nature of the silicon oxygen bond is somewhere in between these two models. In terms of understanding the origin of the piezoelectricity of quartz, a hybrid model results in the silicon atoms being slightly positively charged while the oxygen atoms are slightly negatively charged. In this case piezoelectricity occurs if under pressure the relative displacement of the silicon atoms is either larger or smaller than the relative displacement of the oxygen atoms.

The unit cells of β and α-quartz are compared at the bottom of Fig. 1.7. Each unit cell shows the positions of the silicon atoms (small circles) and oxygen atoms (large circles) that were determined by Lawrence Bragg and R. E. Gibbs. The silicon atoms form two triangles that share a common atom. The oxygen atoms form two separate triangles. It is obvious that β-quartz exhibits a higher symmetry than α-quartz. Because of its symmetry, compressing the unit cell of β-quartz in the horizontal direction does not break the balance between the electric charge on the atoms on one side of the unit cell from the other.

However, in α-quartz the respective triangles of silicon and oxygen atoms are free to rotate independently because there are fewer symmetry restrictions. R. E. Gibbs calculated the rotations and distortions of the oxygen and silicon triangles that result from compression of the unit cell in the horizontal direction in the figure. The center of the negatively charged oxygen triangles are displaced further for a given compression than are the centers of the positively charged silicon triangles. This leads to a net polarization. The piezoelectricity in quartz results from this net displacement on compression of the negatively charged oxygen ions relative to the positively charged silicon atoms. This displacement is possible because of the loss of symmetry that occurs when the structure of quartz changes from the high temperature structure to the low temperature structure. So the origin of piezoelectricity in quartz can be traced to the restrictions that are imposed or not imposed on a crystal as a result of its symmetry.

As described above, quartz (sand) is a hidden component in much of the technology on which we depend like cell phones, computers, watches, sensors, and other electronics. The next chapter explores how nature and the chemical industry take sand and make a wide variety of useful products by changing the structure or architecture of sand.

BOX 1.1 Symmetry and the Structure of Crystals

The proof that crystals are made up of ordered arrays of atoms and the ability to determine where the atoms are in a crystal are triumphs of twentieth century science. In the nineteenth century, mineralogists looked at the external symmetry of the crystals that they found in nature and they wondered about their internal structure.

Mineralogists realized that the faces of crystals could be related to other faces by one of four different types of symmetry operations. Imagine being able to sit inside at the center of a crystal and to look out to the faces that bound the crystal. One type of symmetry operation is inversion, in which every face on the surface of a crystal has an equivalent face on the opposite side of the crystal. This is referred to as a center of symmetry. For example, if one were sitting inside of a cube, then looking right or left, front or back, or up or down looks the same. On the other hand, if one were to sit inside a tetrahedron, then right or left, front or back, and up or down would not look the same. A cube has a center of symmetry and a tetrahedron does not.

A second type of symmetry operation is rotation where a crystal looks the same after rotation around an axis that passes through the center of the crystal. A cube can be rotated by 90, 180, or 270° around an axis that is perpendicular to any face of the cube and that passes through the center of the face. It has a four-fold axis of rotation. Similarly a tetrahedron can be rotated by 120 or 240° around an axis perpendicular to the triangular faces. It has a three-fold axis of rotation.

The third symmetry operation is reflection through a mirror plane. In both the cube and the tetrahedron there are a number of mirror planes in which one half of the cube or tetrahedron is the mirror image of the other half. Finally there are combinations of a rotation followed by inversion.

There are only 32 different possible unique combinations of these symmetry elements, and these are called "point groups." Mineralogists classify crystals by the point group to which they belong. Scientists began to wonder if the external symmetry of a crystal reflected its internal symmetry.

In 1782 Rene Haüy suggested how the external and internal symmetry are related. The story goes that he dropped a crystal of calcite and that it broke into a number of little crystals. Haüy noticed that each of these had the same shape as the original crystal. He concluded that crystals are made up of microscopic building blocks that we now call unit cells (Haüy 1782). These are repeated millions of times in each direction to make a macroscopic crystal. This is like stacking a pile of bricks as illustrated in Fig. 1.8.

To construct a macroscopic crystal, a three-dimensional lattice of points is defined and the points correspond to the corners of all of the unit cells. Starting at one corner of a unit cell, imaginary lines or axes are drawn from the corner along each of the edges of the unit cell. These axes define the lattice as illustrated for a cubic lattice in Fig. 1.8 and for a hexagonal lattice in Fig. 1.7. As envisioned by Haüy, the unit cells are repeated millions of times in all directions to form a macroscopic crystal.

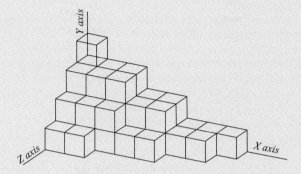

Figure 1.8 Rene Haüy proposed that crystals are composed of small building blocks that are now called "unit cells." They are repeated on a lattice to form a macroscopic crystal.

In 1850, Auguste Bravais, a French physicist and mathematician, proved that exactly 14 distinct lattices can be defined (Bravais 1850) such that the crystal looks the same when viewed from any lattice point. Examples are the cubic lattice in Fig. 1.8 where the sides of the cube are equal in length, and the angles between each of the sides is 90°. A second example is the hexagonal lattice in Fig. 1.7 where the length of one side of the unit cell is different from the other two and the angle between the latter two is 120°. A third example is the face-centered lattice described in Box 2.1.

By the 1890s William Barlow, Yevgraf Fedorov, and Arthur Schönflies independently proved that when the 32 point groups are combined with the 14 Bravais lattices, the result was 230 distinct, independent combinations of symmetry elements involving inversion, rotation, reflection, and translation. These are called space groups, and the structures of almost all crystalline solids can be described by one of the space groups (Barlow 1894; Fedorov 1895; Schönflies 1891). (In 1982, Daniel Shechtman found a class of crystals, quasicrystals, that do not have translational symmetry.)

The classification of the symmetry of a crystal in terms of its space group leads to the remarkable conclusion that only the positions in space of a few atoms have to be defined in the unit cell. The positions of the rest of the atoms in the unit cell will be defined by the symmetry operations of the appropriate space group. In turn the positions of all the atoms in a macroscopic perfect crystal will be obtained by translating the unit cell to every other point on the lattice. For example, by defining the position of one atom

(cont.)

BOX 1.1 *Continued*

of silicon and one atom of oxygen in the unit cell of β-quartz, the positions in space of the more than 10^{23} atoms in a perfect macroscopic crystal can be calculated.

Real crystals have imperfections, and they show deviations from this idealized model. However, as we shall see in Chapter 3, crystals of silicon, which are the foundation of the electronics industry, have few imperfections and approach the model of perfect crystals.

To illustrate how symmetry is used to define the positions of the atoms in the unit cell, consider the oxygen atoms in a unit cell of β-quartz (Fig. 1.9). The space group of β-quartz contains a number of different symmetry operations, but we only need to use two of them to relate all of the oxygen atoms in the unit cell. There are two different types of axes of rotation shown in the figure. One is called a screw axis and is given the designation 3_1. This is an operation that involves two steps. First, all of the atoms in the unit cell are rotated counter clockwise by 120° around an axis perpendicular to the plane of the paper. Then, the rotated atoms are translated by one third of the distance of the unit cell dimension into the plane of the paper. The other symmetry operation is a two-fold axis of rotation in which all of the atoms in the unit cell are rotated by 180°. Once the position of atom 1 is defined, then the two types of axes of rotation will define the positions of all of the remaining oxygen atoms inside the unit cell. This sequence is shown for atoms 1 through 6 in the figure.

Starting from the chance observation of Rene Haüy and the subsequent geometric proofs that established the point groups and the space groups, a mathematical framework was developed by the end of the nineteenth century that allows one to describe the structure of a crystal on the basis of its symmetry.

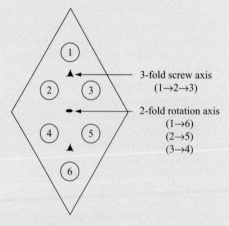

Figure 1.9 The positions of the oxygen atoms in the unit cell of β-quartz are related to one another by two types of axes of rotation as shown on the right. Once the position of atom one is defined, then the positions of the other five atoms are defined by the different axes of rotation.

2
The Architecture of Sand

The great buildings and monuments of the world inspire awe—the pyramids, the great cathedrals, modern museums, and skyscrapers. We marvel at the creativity of the architects and engineers who designed and built them from stones, concrete, and steel. On a different scale we are captivated by nature's architecture in gemstones and minerals. We appreciate the flash of light, depth of color, and numerous forms found in quartz gemstones and minerals that nature itself makes from sand. On the atomic scale, scientists have come to understand the architecture of everything from gems and minerals to proteins.

In the twentieth century, scientists learned how to control the architecture of the atomic world in order to synthesize new materials to meet the evolving needs of technology. The architects of the atomic-scale world build structures with varying degrees of open space and lightness of structure. Not unlike the architects of present day buildings and bridges, the architects of the atomic-scale world are testing the strength and stability of the new structures that they have synthesized in the laboratory.

This chapter looks at how the architects of the atomic-scale world take sand and make a wide variety of new materials that have become everyday products in our lives. Before looking at these new materials let us start by asking how scientists determine the architecture of sand? As discussed in Chapter 1, the concepts of symmetry were enumerated in the nineteenth century, and a mathematical framework was developed to relate the positions of the atoms in a crystal by the symmetry operations of the appropriate space group. However, there was no way to determine where the atoms were in the unit cell or to which space group a crystal belonged. Furthermore, there was no direct way to determine the distance between the atoms in a crystal and the dimensions of the unit cell. None the less, by the end of the nineteenth century scientists began to speculate what the architecture of the atomic world might look like.

One of these was William Barlow, who was an amateur geologist. He had inherited a sizable fortune and was able to devote his time to studying the symmetry and structure of crystals. W. J. Pope writes in Barlow's obituary:

"Barlow thus found himself in his early thirties with an independence, with a genius for handling geometrical problems of a particular kind, and with ample leisure to devote to the study of crystal structure, which had become the subject of his choice. He had not, however, received that rigid disciplinary training through which most students of physics and chemistry acquire a broad sense of contemporary knowledge of the physical universe, In some respects this was a hindrance but in others an advantage; it left a powerful intellect unhampered by authority and led a logical mind to pursue its inquiries into difficult and obscure paths which might intimidate the more conventionally trained." (http://pubs.rsc.org/en/content/pdf/article/1935/jr/jr9350001327)

One of his major contributions was the derivation by novel geometric arguments of the 230 space groups. Then, he proposed the crystal structures of a number of materials. In 1883, Barlow showed that there are two ways to close-pack spheres to most efficiently fill space and, in fact, a large number of metals have one of these two arrangements. In 1898 he proposed crystal structures for simple salts like NaCl and CsCl. The structures were based on the close-packing of spherical atoms with different sizes (Barlow 1898). At the time his ideas were met with skepticism, but as we shall see he was correct.

It became increasingly important to develop a technique by which the crystal structure of a material could be determined in the laboratory. The story of this achievement begins with the discovery of x-rays in 1895 by Wilhelm Röntgen, a physics professor at the University of Würzburg. Like many scientists at the time he was studying the properties of electrons, then called cathode rays. To generate electrons, metal electrodes are placed at each end of a partially evacuated glass tube. When a large electric field is discharged across the electrodes some of the remaining gas molecules in the tube are ionized. The resulting positively charged ions are attracted to the negative electrode, the cathode, and when they hit the cathode, large numbers of electrons are produced. These in turn are accelerated toward the positive electrode, the anode. To observe the cathode rays, the end of the tube is coated with a fluorescent material. Many of the rapidly moving electrons pass by the anode and hit the end of the tube producing a glow. Röntgen had blocked the glow from the coating with cardboard. Quite unexpectedly, he observed a faint glow on a sheet of fluorescent material that was lying a few feet away from the tube. The appearance of the glow was correlated with the electrical discharge in the tube. He proceeded to show that the origin of this new kind of radiation was the anode. The electrons hitting the anode had caused a new type of radiation to be emitted.

This new kind of ray passed right through the glass tube, whereas electrons were absorbed in the glass. The penetrating power of this new radiation is illustrated by the famous picture of the hand of Röntgen's wife with the clear outline of all of the bones in her hand and her wedding ring. Röntgen gave this new type of radiation the name x-rays to indicate that their nature was unknown. Wilhelm Röntgen was awarded the first Nobel Prize in Physics in 1901 for the discovery of x-rays.

The seminal experiment to understand the nature of x-rays was performed at the University of Munich in 1912 by Max von Laue, Paul Friedrich, and Walter Knipping (Friedrich *et al.* 1912). This experiment completely revolutionized our understanding of both the nature of x-rays and of the structure of crystalline materials. It was known at the beginning of the century that light and sound propagated as waves, but von Laue put forth the idea that Röntgen rays or x-rays also propagated like waves. Further, he speculated that the spacing between atoms is comparable to the wavelength of the x-rays. If the atoms in a crystal are arranged in periodic arrays, then the x-rays that scatter off the atoms interfere with each other. Just as light is diffracted from a diffraction grating, x-rays are diffracted by crystals. Von Laue asked Friedrich and Knipping to try to observe the diffraction of x-rays.

They did an experiment in which a beam of x-rays was directed toward a photographic plate. They placed a crystal (they chose to use copper sulfate and zinc sulfide) between the x-ray source and the photographic plate. When the plate was developed, it showed a symmetric pattern of spots arrayed around the spot created by the x-ray beam. The x-rays were diffracted by the crystal to form the symmetric pattern.

Von Laue derived a set of three equations (now known as the Laue equations) that both described the diffraction from a three-dimensional array or lattice of atoms and accounted for the positions of the spots on the film. Von Laue, Friedrich, and Knipping had established that x-rays propagated as waves and that crystals were composed of ordered arrays of atoms with the spacing between the atoms being comparable to the wavelength of the x-rays. Max von Laue was awarded the Nobel Prize in Physics in 1914.

Word of the discovery of the diffraction of x-rays spread rapidly across Europe, and it was a twenty-two-year-old graduate student in physics at Cambridge University, William Lawrence Bragg, who was in the right place at the right time to take the next step. Max von Laue had shown that the atoms in copper sulfate and zinc sulfide were arranged in periodic arrays, but what was the actual arrangement of the atoms in the unit cell, in other words, what was the crystal structure? Bragg determined the crystal structure of zinc sulfide and became the father of x-ray crystallography.

Lawrence Bragg had received copies of the papers of von Laue, Friedrick, and Knipping from his father, William Henry Bragg, who was a professor of physics at the University of Leeds. (Both father and son had the same first name, William, so the son is referred to by his middle name—William Bragg, the father, and Lawrence Bragg, the son.) As a student, Lawrence Bragg was taking courses on optics and learning how light waves were reflected and refracted from crystals. At the same time, he was aware of the predictions for the structures for simple compounds that had been made by William Barlow (Hunter 2004).

Lawrence Bragg had the insight to use ideas from optics and to make the analogy between the reflection of light and the reflection of x-rays from planes of atoms in a crystal. As described in Box 2.1, this leads to an equation that relates the distance between the planes to the wavelength of the x-rays and the angle of diffraction. This is now known as Bragg's Law, and it is the basis for all x-ray crystallography. Using Bragg's Law and Barlow's ideas about the most efficient way to pack atoms in a crystal, Lawrence Bragg proposed a model for the crystal structure of zinc sulfide that accounted for both the position and the relative intensity of all of the diffraction spots observed by Friedrich and Knipping (Bragg 1913 and Bragg and Bragg 1913).

One of the most successful father and son collaborations in science grew out of this experiment (Jenkin 2008). William Henry Bragg, the father of William Lawrence Bragg, was born on 2 July 1862 in Cumberland England between the Solvay Firth and the Irish Sea. He went to Trinity College, Cambridge in 1881 and, as was common at that time, started his scientific career at one of the universities in the broad reaches of the British Empire. In 1886 he became professor of mathematics and experimental physics at the University of Adelaide in Australia, where his first few years were devoted to teaching and to establishing a strong physics department. His son, William Lawrence Bragg, was born in Adelaide on 31 March 1890. After the discovery of x-rays in 1896, William set up an x-ray tube and took the first medical x-ray in Australia of his son's broken arm.

As with the Curies, William Bragg became interested in Becquerel rays. The Curies had concentrated on looking for other materials that exhibited this new type of radiation, and William Bragg began to study how these rays were absorbed by different materials. The last decade of the nineteenth century was an exciting time in science with the discovery of the electron by J. J. Thompson (1895), x-rays by Wilhelm Röntgen (1896), and alpha particles by Henri Becquerel (1896). As beams of electrons and x-rays pass through materials, they are absorbed, and the intensity of the beams decreases exponentially with increasing depth. William Bragg found that alpha particles behaved differently. They were mainly absorbed at a definite depth inside the material rather than decreasing steadily with increasing depth. (This result will be discussed further in Chapter 4.) These results and experiments on the nature of x-rays led to William's growing reputation in the scientific community and, in 1909, he was offered the position of Cavendish professor of physics at the University of Leeds in England.

At Leeds, William turned his attention to the properties of x-rays. Charles Barkla, a British physicist at the University of Edinburgh, had studied the absorption of x-rays. He found that when a beam of x-rays is absorbed in a material that x-rays with a different, longer wavelength are re-emitted. The re-emitted x-rays depended on the chemical elements in the material, and each element had a set of characteristic x-rays. The fact that each element, independent of its chemical environment, had a unique signature in its pattern of characteristic x-rays gave chemists a new type of spectroscopy with which to analyze unknown materials. Charles Barkla received the Nobel Prize in Physics for this discovery in 1917.

William Bragg built the first x-ray diffractometer to accurately determine the wavelengths of the characteristic x-rays. Bragg's Law relates the spacing between the planes in a crystal to the wavelength of the x-rays. If either quantity is known, then the other can be calculated from Bragg's Law and the angle of diffraction. In a diffractometer a beam of characteristic x-rays emitted by an element is directed toward a crystal. The angle that the face (planes) of the crystal makes with the beam of x-rays is slowly increased, and the angle that the x-ray detector makes with the x-ray beam is increased at twice that rate. This is similar to the reflection of light from a mirror. The angle at which the incoming beam hits the crystal is equal to the angle at which the outgoing beam leaves the crystal. A large increase in intensity of scattered x-rays is observed at the angle where Bragg's Law is satisfied for the characteristic x-rays of the element used as the source of x-rays. If one uses the same crystal but different elements as the source of x-rays, then one can accurately measure the differences in the characteristic x-rays for each source. William Bragg used the diffractometer to determine the wavelengths of the characteristic x-rays of the elements iridium, nickel, platinum, and tungsten.

An interesting sidebar to William Bragg's research on characteristic x-rays illuminates how physics priorities were set at the beginning of the twentieth century. Probably the most famous professor of physics in England at the time was Ernest Rutherford, who we will discuss in Chapter 4. Rutherford had begun his career in New Zealand and had been a mentor and friend of William Bragg. Rutherford, who was then at the University of Manchester, had two students, C. G. Darwin and G. J. Mosley, who were also studying the characteristic x-rays of the elements. Bragg had in fact tutored them on the use of the diffractometer. Rutherford asked Bragg to delay publication of his results until Darwin and Mosley could get their own results ready for publication, and Bragg complied with his friend's request. This type of gentleman's agreement on priorities in doing research has greatly diminished in recent times owing in part to the competitive nature of scientific research and to the pressures of obtaining research funds. Bragg decided to leave that area of research to them and to collaborate with his son on the structure of materials (Jenkin 2008).

Over the next couple of years they determined the crystal structures of 16 different materials. They began with the simple crystals, like sodium chloride and potassium chloride and diamond, which is a form of the element carbon. (The structure of silicon, which is the same as that of diamond, is discussed in Box 2.1.) They continued on to more complicated minerals, like quartz that was described in Chapter 1.

This new tool to determine the architecture of the atomic world changed many prevailing ideas of the structure of materials. For example, chemists had been convinced that crystals of sodium chloride were made up of molecules of "NaCl." They reasoned that molecules are held together by chemical bonds. Sodium has one bonding electron and chlorine needs an additional electron to fill its outer shell of electrons. By bonding together, both sodium and chlorine would have filled outer shells of electrons. However, the structure determined by x-ray diffraction showed that each sodium atom was surrounded by six chlorine atoms that were equally spaced, and each chlorine was surrounded by six equally spaced sodium atoms. There were no molecules of "NaCl" in a crystal of sodium chloride. Actually knowing how the atoms are arranged in salt in turn led to the concept of ionic bonding that was developed by Linus Pauling and mentioned in Chapter 1.

Unfortunately, the onset of the First World War in 1914 temporarily stopped the development of x-ray diffraction. Both Braggs contributed to the war effort. They used their knowledge of physics to develop systems to detect submarines and gun emplacements. William Bragg worked for the British Board of Invention and Research. He developed a directional hydrophone to listen for submarines but, as discussed in Chapter 1, the echo sonar developed by Paul Langevin became the preferred method of submarine detection.

Lawrence Bragg was in the artillery and developed a sound ranging system to locate enemy gun emplacements. He set up a series of microphones spaced along the trenches and recorded the differences in the time of arrival of the report of an enemy cannon being fired. From those differences, he could calculate the position of the German cannon. He could also listen for the reports of British shells exploding near to the German battery and tell the British artillery how to adjust their fire to destroy it.

In 1915, while at the front in France, Lawrence received a letter that announced that he and his father had been awarded the 1915 Nobel Prize in Physics for their work on x-ray spectroscopy and x-ray diffraction. Lawrence Bragg became and still is the youngest scientist to win the Nobel Prize. Because of the war, the prizes were not presented until 1920.

William Bragg moved to University College London in 1915 and, after the war, started the first effort to determine the structures of organic molecules like anthracene and naphthalene. One of his students, Kathleen Lonsdale, determined the structure of benzene and went on to become the first woman elected a Fellow of the Royal Society and the first female tenured professor at University College London.

Lawrence Bragg moved to the University of Manchester after the war and continued research using the diffraction of x-rays to determine more and more complicated crystal structures (Box 2.1). He also used x-rays to determine the size of atoms in a crystal and to study subtle changes in the structure of materials as a function of temperature. A well known model for a particular type (order–disorder) of transition is called the Bragg–Williams model.

In 1938 he became director of the Cavendish Laboratory at Cambridge University. The position was held by Ernest Rutherford until his death, and it was quite an honor for Lawrence Bragg to be appointed to the most prestigious position in physics in England. During his tenure he encouraged the development of new areas of research such as biology. He supported a young graduate student named Max Perutz who wanted to pursue what seemed at the time an impossible goal—to solve the crystal structure of hemoglobin. This was long before computers, and the diffraction pattern of a large molecule like hemoglobin contains thousands of Bragg reflections. Perutz had to measure the intensity of a large fraction of them. To determine where the atoms are in the unit cell of a crystal of hemoglobin, he had to make thousands and thousands of calculations using mechanical calculators. This was a tour de force and in the end through enormous perseverance Perutz determined the structure.

It was while Lawrence Bragg was the director of the Cavendish Laboratory that John Kendrew, Francis Crick, and James Watson joined the laboratory. Crick and Watson shared the Nobel Prize in Medicine with Maurice Wilkins in 1962 for the discovery of the structure of DNA, and Perutz and Kendrew shared the Nobel Prize in Chemistry in the same year for determining the structure of globular proteins. Sir William Lawrence Bragg determined the first crystal structure, zinc blende, thereby founding the field of x-ray crystallography, and near the end of his long and distinguished career, he directed the laboratory that determined the structure of life itself.

BOX 2.1 X-ray Diffraction and the Structure of Crystals

When we go to the supermarket, everything has a bar code, a one-dimensional sequence of bars, stamped on it for identification. It is the way that the laser scanner at the checkout counter recognizes what is being bought. In biology and in law enforcement, DNA sequencing produces an analogous bar code, which can be used as a means of identification. Each bar represents a different DNA fragment that can be identified by its position on the strip. The pattern of bars for each sample is unique.

The bar code of materials science is x-ray powder diffraction in which the intensity of the x-rays diffracted from a sample composed of millions of randomly oriented microcrystals (referred to as a powder) is recorded as a function of the

(*cont.*)

BOX 2.1 *Continued*

diffraction angle. This gives a one-dimensional sequence of Bragg reflections. When a new material is synthesized, an x-ray powder diffraction pattern is obtained to provide a unique characterization of the material.

The fundamental idea behind x-ray diffraction is the interference of waves. Lawrence Bragg considered planes of atoms and how waves scattered from adjacent parallel planes interfere. The geometric construction that illustrates the diffraction of x-rays is shown in Fig. 2.1. X-rays with a wavelength "λ" scatter off a set of parallel planes that are separated by a distance "d." The figure shows the resulting wave scattered from each plane.

There is a unique angle, the Bragg angle, at which the wave scattered from the second plane is exactly one wavelength behind the wave scattered from the first plane as illustrated in the figure. As a result, the peaks and valleys of both scattered waves coincide, and they are said to be in phase. The scattered wave from each successive plane in a crystal will be one more wavelength behind, and the amplitude of the scattered x-rays will increase with the number of planes. For other angles, the scattering from different planes will be out of phase, and no enhanced scattering will result because some waves will cancel other waves. The mathematical relation between the wavelength, the spacing between planes and the angle that leads to all the scattered x-rays being in phase is called Bragg's Law.

As discussed in Chapter 1, the structure of a crystal is described by the positions of the atoms inside a building block called the unit cell. The structure of the element silicon is shown in Fig. 2.2. It has the same structure as diamond, and the unit cell is a cube that contains eight silicon atoms (atoms on the face of the unit cell are shared by two adjacent unit cells and count as one half of an atom. Similarly, the atoms at the corners are shared by eight unit cells and count as one eighth of an atom).

When the unit cell is rotated, there are directions in which there are planes of atoms similar to those discussed above in the description of Bragg's Law. In

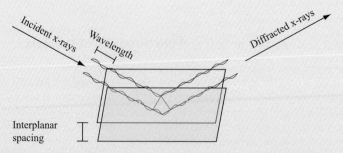

Figure 2.1 X-ray diffraction is based on Bragg's Law. At the Bragg angle, x-rays diffracting from the bottom plane travel exactly one wavelength further than those from the top plane. The diffracted x-rays add constructively. At other angles the diffracted waves interfere thereby reducing the intensity.

Fig. 2.2 planes that are perpendicular to three different directions in the cube are shown. These include the body diagonal, the face diagonal, and the cube edge.

To determine the crystal structure of a material, the intensity of the diffracted x-rays is measured at the diffraction angles corresponding to as many planes as possible. In a modern version of William Bragg's diffractometer, a computer controls stepping motors that orient both the crystal and the x-ray detector to sequentially measure the intensity of the scattering from the different planes.

For less complicated crystals a collection of microcrystals with random orientation (a polycrystalline or powder sample) can be used. With increasing angle, a series of Bragg reflections will be observed when different planes of atoms in the structure satisfy Bragg's Law for some of the microcrystals. This is referred to as an x-ray powder diffraction pattern. The x-ray powder diffraction patterns (bar codes) for the element silicon and for quartz (sand) are shown in Fig. 2.3.

From the measured diffraction angles, a set of interplanar spacings or "d" spacings is calculated using Bragg's Law. The crystallographer in turn deduces the dimensions and the symmetry of the unit cell that uniquely accounts for all the "d" spacings. The space group and the positions of the atoms in the unit cell are determined by analyzing the relative intensities of the different Bragg reflections.

For example, in Fig. 2.2 the spacing marked "d_4" corresponds to planes of atoms that are separated by one fourth of the unit cell dimension, a/4, where "a" is the dimension of the unit cell. Mathematically there are also planes that are separated by a, and a/2. However, for planes that are separated by a/2 there are

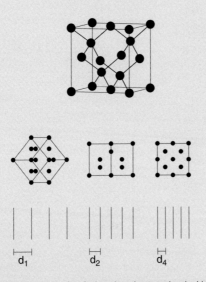

Figure 2.2 The structure of silicon (top) showing the tetrahedral bonds around each of the silicon atoms inside the unit cell. The structure of silicon is projected perpendicular to three directions in the unit cell (body diagonal, face diagonal, and cube edge) to illustrate the spacing between the different planes of atoms that are shown at the bottom.

(*cont.*)

BOX 2.1 *Continued*

planes of atoms that are halfway in between those planes. X-rays scattered from these planes are exactly one half a wavelength behind the atoms in the planes that are separated by a/2. Their peaks and valleys are exactly opposite and cancel the scattering from the planes of atoms that are separated by a/2. As a result, the intensity of a Bragg reflection corresponding to a spacing of a/2 is zero. A similar analysis shows that the intensity of a Bragg reflection corresponding to planes separated by a distance "a" is also zero.

To determine the crystal structure of diamond, the Braggs systematically analyzed all possible Bragg reflections for a cubic unit cell to see which ones had a measurable intensity and which ones had zero intensity. This is a mathematical puzzle and a unique solution is found by trial and error. Today, sophisticated computer programs sort through possible solutions to the puzzle and find the best agreement between the observed Bragg reflections and those calculated for different models. The Braggs concluded that the structure of diamond had a face-centered cubic Bravais lattice.

The position of an atom is defined in terms of three coordinates x, y, and z that give the position in terms of the fraction of the unit cell dimension along the three axis of the unit cell. For a face-centered cubic Bravais lattice, if there is an atom

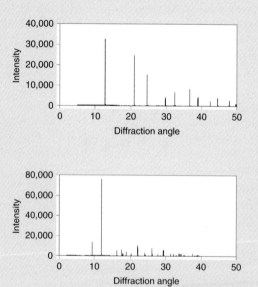

Figure 2.3 X-ray powder diffraction patterns (bar codes) for silicon (top) and α-quartz (bottom). The 1st, 2nd, and 4th Bragg reflections for silicon correspond to diffraction from the planes shown in Fig. 2.2. The unit cell is derived from the angles of the Bragg reflections using Bragg's Law, and the positions of the atoms inside the unit cell are derived from the intensities of the Bragg reflections. (Courtesy of Peter Stephens, State University of New York at Stonybrook. Measurements done at the National Synchrotron Light Source at Brookhaven National Laboratory that is supported by the US Department of Energy.)

at (x, y, z), then symmetry requires that there will also be atoms at (½+x, ½+y, 0), (½+x, 0, ½+z), and (0, ½+y, ½+z). After much trial and error the Braggs found that the structure of diamond had an atom at (0, 0, 0) and those generated by the face-centered lattice (½, ½, 0), (½, 0 ½), and (0, ½, ½). There was also an atom at (¼, ¼, ¼) and the face-centered cubic positions (¾, ¾, ¼), (¾, ¼, ¾), and (¼, ¾, ¾) (see Fig. 2.2 for silicon). This model for the crystal structure of diamond accounted for all of their observations. The model predicted those Bragg reflections that have measurable intensities and those that have zero intensity.

In the structure of diamond or silicon all of the atoms are at positions corresponding to simple fractions of the unit cell dimensions, and these are called special positions. This was also true for many of the other structures that the Braggs determined in their early work like zinc blende (ZnS), NaCl, and KCl. However, the structure of quartz is more complicated as is evident from the powder pattern in Fig. 2.2. The structure of quartz (Fig. 1.7) has atoms that do not occupy positions that are all simple fractions of the unit cell dimensions. The atoms no longer lie exactly on simple planes so the scattering from individual atoms in a unit cell may either add or subtract from the overall scattering from a given plane. To solve the structure of quartz, Lawrence Bragg and R. E. Gibbs had to calculate how the intensity of each observed Bragg reflection varied for different values for (x,y,z) for both the silicon and the oxygen atoms. After many attempts they found a set of values that were consistent with all of the measurements. This is the structure described in Chapter 1.

As crystallographers tried to solve the structures of increasingly more complicated materials, it was obvious that this type of trial and error search for a structure that uniquely accounted for the diffraction measurements was becoming more and more difficult. Again, with his knowledge of optics, Lawrence Bragg realized that Fourier analysis could be used to determine crystal structures (an illustration of Fourier analysis is at: http://www.ysbl.york.ac.uk/~cowtan/fourier/ftheory.html). A mathematical operation called a Fourier transform is used to convert the set of amplitudes for all the Bragg reflections into a three-dimensional map of the electron density in the unit cell. As x-rays scatter from the electrons in an atom, the map shows directly the positions of the atoms in the structure where there is a large concentration of electrons.

The problem is that the intensity is measured in an x-ray diffraction experiment, and the intensity is the square of the amplitude. A wave is characterized by an amplitude and a phase that measures how far the wave has advanced at any given instant from an arbitrary point. Both the amplitude and the phase are needed to calculate the Fourier transform. The history of the field of x-ray crystallography since the 1920s centers on trying to find ways to determine the phase of the scattering from each Bragg reflection.

In the first application of Fourier analysis, the structure of the mineral diopside $CaMg(SiO_3)_2$ was solved by Lawrence Bragg and Bertrum Warren in 1926. They calculated a one-dimensional Fourier projection along one of the principle directions of the unit cell. This required making 35,000 calculations by hand!

Various mechanical aids and other types of projection functions were developed in the 1930s and 1940s to facilitate the determination of the structures of complex materials with many independent atoms and ever increasing unit cell dimensions. In retrospect, a remarkable number of complex crystal structures

(cont.)

BOX 2.1 *Continued*

were solved by crystallographers in the 1930s, 1940s, and 1950s based to a large extent on their intuition and knowledge of structural chemistry.

One example that I have always admired is the determination of the structure of the alpha phase of plutonium metal during the Manhattan Project in the Second World War (Zachariasen and Ellinger 1963). In order to understand the chemistry of these new manmade elements, the crystal structures of the transuranium elements and their chemical compounds had to be determined. As plutonium was made literally atom by atom in a nuclear reactor, only small polycrystalline samples were available in the beginning, and W. H. Zachariasen solved the structure from a powder diffraction pattern. Because of the complexity of the structure, the intensities of many of the Bragg reflections that occur at low angles were not observable. Normally one relies on the Bragg reflections that occur at low angles to determine the unit cell. Zachariasen was able to determine the monoclinic unit cell of alpha plutonium and in turn its unusual structure. In a monoclinic crystal all three unit cell dimensions are different and one of the angles between the axes is not 90°. My first scientific paper in 1961 on the very much simpler structure of americium metal, the next element after plutonium in the periodic table, was in collaboration with Zachariasen, and it was always a humbling experience to watch him solve crystal structures by trial and error.

Needless to say, it was the development of the computer that opened the door to solving the structures of really complex materials like proteins. Not only did this enable the calculation of electron density once the phases had been determined, but it allowed crystallographers to optimize the positional parameters of all of the atoms in the unit cell. The crystallographer proposes a model for the structure of a material and then calculates the amplitudes of the Bragg reflections expected for that model. The calculated amplitudes are then compared with the measured amplitudes. Using high speed computers, the crystallographer minimizes the differences between the observed and calculated amplitudes to obtain the most accurate model of the structure.

2012 is the hundredth anniversary of the discovery of x-ray diffraction. It has revealed the architecture of the atomic world from simple compounds to complex biological materials, and it has proved to be one of the most important scientific discoveries of the twentieth century.

To illustrate the importance of x-ray crystallography to twentieth century science, let's look at a group of materials derived from sand (Fig. 2.4) to see how scientists modify the architecture of sand to produce new materials. The materials in Fig. 2.4 have obvious visual differences, but what is less obvious is that they have an enormous range of densities. The densities differ by a factor of 100 from 5 gm/cm^3 for the quartz in the flint arrowhead to 0.03 gm/cm^3 for the aerogel.

It is hard to appreciate from a black and white illustration what a density approximately 100 times lower than that of quartz means. The density of aerogel is only approximately 30 times that of air! Pick up a piece of this translucent blue material and it appears to have almost no

Figure 2.4 Materials made from sand (silica): (top) (a) flint arrowhead and (b) quartz optical fiber (reprinted with permission of Alcatel-Lucent USA Inc.); (middle) (c) scolecite (courtesy of the President and Fellows of Harvard College) and (d) opal (courtesy of the Smithsonian Museum); (bottom) (e) silica gel and (f) aerogel. The networks of corner-sharing tetrahedrons on going from flint to aerogel are progressively more open with a corresponding decrease in density by a factor of a hundred.

weight at all. In Fig. 2.4, the aerogel is sitting on a bed of sand, and one can see the sand through the semitransparent aerogel.

The architecture of sand, as revealed by x-ray diffraction, is constructed from one building block, the tetrahedron (Fig. 2.5). The chemical properties of silicon and oxygen result in an architecture in which there is a network of tetrahedrons. Each tetrahedron is composed of a silicon atom at the center and oxygen atoms at the four corners. The tetrahedrons are linked together by sharing corners to form extended networks that vary depending on the material.

The tetrahedrons occur because of the directional chemical bonds that are the glue that hold atoms together. The model of directional chemical bonds was developed in the 1920s by Gilbert Newton Lewis and Linus Pauling in the early days of quantum mechanics, as described in the last chapter. The electrons in an atom occupy orbitals that define the probability of finding an electron in terms of its distance from the nucleus and in terms of the symmetry of the orbital (Pauling 1960). The lobes of the orbitals in silicon point toward the corners of a tetrahedron. In sand, the electrons in the four lobes form bonds with four adjacent oxygen atoms (Fig. 2.5). Each oxygen in turn is shared by two adjacent tetrahedrons leading to a three-dimensional network of corner-sharing tetrahedrons.

The materials in Fig. 2.4 differ not only in the density of the network of tetrahedrons but also in the degree of order or regularity in the network. The crystal structure of quartz was discussed in the last chapter (Fig. 1.7). Quartz crystals have an ordered array of tetrahedrons, but the quartz fiber in Fig. 2.4 has a random network of tetrahedrons. Flint and zeolites like scolecite are crystalline (or composed of microcrystallites) and have ordered networks. Opal has a unique partially ordered structure, and silica gels and aerogels are disordered and have random networks.

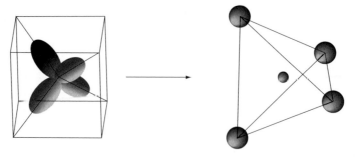

Figure 2.5 The electron orbitals in the silicon atom point toward the corners of a tetrahedron, and the silicon atom forms bonds with four oxygen atoms. As a result, the architecture of sand can be viewed as a three-dimensional network of corner-sharing tetrahedrons.

The most common materials with a random network of tetrahedrons are the many forms of glass. They are made by heating together silica and two or more metal oxides. The nature of a glass depends on the mixture, but glass is usually hard and brittle. When it is heated, it softens and can be blown, rolled, or shaped into any desired form. The glass industry uses about 10 million tons of sand yearly to make glass containers, windows, fiberglass, and a whole host of everyday products. There are many books on the history and uses of glass, and we defer to them for more extensive discussions (Macfarlane and Martin 2002). Instead, our discussion of sand concentrates on comparing the structure of crystals with the structure of disordered materials like glass. When silicon dioxide is melted and then the liquid is cooled rapidly, it forms a glass rather than one of the crystalline forms of quartz. The structure is still composed of corner-sharing silicon–oxygen tetrahedrons, but they form a random network. In a glass only average distances between the atoms and the average angles between different chemical bonds can be determined using x-ray diffraction.

The difference between an ordered network of corner-sharing tetrahedrons and a disordered network is illustrated in Fig. 2.6. The drawing on the left is a projection of a layer of the structure of the mineral tridymite that is one of the phases of silica (sand). The tetrahedrons form rings composed of six tetrahedrons that are arranged in an ordered hexagonal array.

On the right of Fig. 2.6 is a hypothetical model of a disordered two-dimensional array of tetrahedrons that is based on one of the first

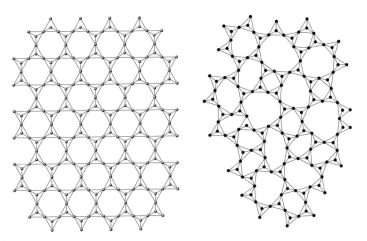

Figure 2.6 Comparison of an ordered network of tetrahedrons in a crystal of the mineral tridymite and a hypothetical, two-dimensional, disordered network in a glass (the latter is adapted from Zachariasen 1932). In tridymite there are ordered rings composed of six tetrahedrons whereas in the model for a glass there are disordered rings of 4, 5, 6, 7 tetrahedrons.

attempts to describe the structure of a glass by W. H. Zachariasen (Zachariasen 1932). In crystalline quartz there are ordered arrays of rings formed by six tetrahedrons. In a random network there will be a mixture of rings that have 4, 5, 6, or 7 tetrahedrons that are all interconnected. In a real glass there is a three-dimensional random array of rings, but the concept is more easily illustrated with the two-dimensional model shown in Fig. 2.6.

Let's follow the changes in the architecture of the tetrahedral networks going from the most dense material—flint—to the least dense—aerogel. Flint was one of the earliest materials used by man to make tools. Pieces of flint aren't like large single crystals of quartz with regular faces (Fig. 1.1). Instead they have irregular shapes. On a microscopic scale flint is composed of millions and millions of microcrystals. X-ray diffraction studies reveal that flint is composed predominantly of microcrystals of quartz. So on the microscopic scale, the structure and the composition of the crystallites is identical to the large crystals of quartz discussed in Chapter 1. Recent studies have found that flint usually contains a small percentage of crystallites of another form of silica called Moganite (Heaney and Post 1992) in addition to quartz.

Flint is found in sedimentary rock formations. Eons ago dissolved carbonates like $CaCO_3$ reacted with SiO_2 to form $CaSiO_3$ and CO_2. The $CaSiO_3$ dissolved in water to form silicic acid, $[SiO_x(OH)_{4-2x}]_n$. (This reaction is important in the growth of marine organisms like diatoms, as discussed later in the chapter. It is also an important chemical reaction in the carbon dioxide cycle that controls the concentration of CO_2 in the ocean.) Over time the silicic acid that is trapped in pockets in rocks became more concentrated as the water evaporated and microcrystals of quartz grew out of the supersaturated solution. The microcrystals were pressed together on geologic time scales to form flint by the weight of the sedimentary layers that deposited on top of the rock formations.

Flint can be shaped into arrowheads and other primitive weapons by a process called knapping. Repeatedly hitting a piece of flint with a stone causes chips to flake off. This allowed early man to shape tools like the flint arrowhead in Fig. 2.4. Later it was found that sparks result when flint is struck against a piece of metal. Small pieces of metal fly off and in the process they are heated by abrasion producing sparks. This method is used to start campfires; to ignite the gunpowder in a flintlock rifle; or to provide the sparks in cigarette lighters and gas barbeques. So, even from prehistoric times materials derived from sand were essential to human progress (Mason 1978).

Many important industrial materials are composed of ensembles of microcrystallites. Mixtures of oxides that have been heated so that the microcrystals adhere to each other are called ceramics. Unlike flint, where the microcrystallites are just pressed together, the heating or sintering of the microcrystallites in ceramics leads to crystallites that are bonded together. Early man learned how to make slurries of clay and then shape and heat them to make pottery. Through the centuries artisans learned how to improve the quality and strength of these materials by controlling the size of the microcrystals and how they were sintered. With the development of x-ray diffraction, twentieth century ceramic engineers correlated the microstructure of the ceramic with its properties to tune the properties according to the application.

Flint and ceramics are composed of microcrystals, in contrast to a glass that is disordered and superficially resembles a frozen liquid. Twentieth century science has led to more complex materials that combine the attributes of microcrystals and glass. These are called glass ceramics, and they are composed of microcrystals embedded in a glass matrix. Some compositions of glass ceramics have very small coefficients of thermal expansion. This means that they can experience rapid changes in temperature without cracking. Because of their insensitivity to thermal shocks, these materials are used to make stovetops like CERAN® by Schott Corporation and cookware such as CorningWare®, CORELLE® (Fig. 2.7), and VISIONS® by World Kitchens, LLC.

The beauty of glass ceramics is that they can be shaped while in the glassy state and then processed to make the composite of microcrystals and glass. To make a glass ceramic, sand and other oxides are heated together to form a glass. The glass is first heated to soften it and then shaped or molded into the desired product. Finally, the glass is heated to a temperature where microcrystals start to grow. The final product is composed predominantly of microcrystals, which have the same structure as phases of silica like β-quartz or cristobalite, embedded in a matrix of the remaining glass.

Figure 2.7 Glass ceramics have low thermal expansion and are used in heat resistant dinnerware such as CORELLE®. They are a composite of silica microcrystals embedded in a glass matrix.

The trick of course is to get the crystallites to grow in the glass so that crystallites of similar size are more or less uniformly distributed in the glass matrix. In order to grow crystallites, small regions or nuclei have to separate out of the surrounding glass. These nucleation sites then proceed to grow to form microcrystals.

In 1956 Stanley Stookey at the Corning Glass Works discovered that the addition of small amounts of titanium oxide to the glass provided the necessary nucleation sites, and this gave birth to the wide variety of heat resistant surfaces and dinnerware that have become another example of everyday products that started with small grains of sand and other oxides (Stookey 1956).

Glass, flint, and ceramics are materials where the microstructure is changed during growth, but the overall density is similar. Let's go back to crystals like quartz and ask: how can the density be changed by modifying the network of tetrahedrons at the atomic scale? The structure of quartz contains tunnels running through the structure as illustrated in Fig. 1.7. There are also tunnels running through other phases of silica. In the mineral tridymite (Fig. 2.6) the tunnels are formed by rings of six corner-sharing tetrahedrons with alternate tetrahedrons pointing up and down. The rings are interconnected to form sheets of six-membered rings.

The minerals quartz, crystobalite, and tridymite have the same composition as silicon dioxide. All three have crystal structures with tunnels composed of six-membered helices (quartz) or rings (crystobalite and tridymite). The difference between the structures of the minerals crystobalite and tridymite is in the way in which the layers are stacked. In both cases the resulting structures have tunnels running through them because the six-membered rings line up on top of each other (Heaney 1994).

The tunnels in quartz, cristobalite, and tridymite are 0.1 nanometers (nm) in diameter, and only very small ions like lithium can fit inside the tunnels. Nature and man make materials with much larger tunnels and correspondingly lower densities. These include zeolites where the tunnels are approximately 1 nm in diameter; silica gels with pore sizes between 2 and 50 nm and aerogels with pores greater than 50 nm (Everett 1972 and Brinker 1996). It is this enormous range in pore sizes that accounts for the large range in densities exhibited by the materials in Fig. 2.4. Let us not forget that we are discussing the structure of materials in the nano-world. To keep the scale of the size of the pores in perspective, the diameter of a human hair is of the order of 100,000 nm.

Zeolite is the name given to a group of naturally occurring minerals and synthetic crystals made from mixtures of silicon and aluminum oxide. They have large tunnels built into a network of tetrahedrons and, as a result, much larger molecules are incorporated inside the tunnels. The Swedish chemist Axel Cronstedt, who first reported zeolites in 1756, was amazed to observe that water bubbled out of them when they were heated. Consequently, he gave them the name zeolites, which means boiling stones.

Zeolites have rigid frameworks of corner-sharing tetrahedrons that have varying ratios of silicon and aluminum ions at their centers depending on the composition. Because aluminum has one less positive charge in its nucleus than silicon, extra positively charged ions occupy the tunnels in the structure to provide charge neutrality. By varying the ratio of silicon to aluminum, the number of ions in the tunnels can be tuned. In addition, the size of the tunnels is controlled by the size of the ions that are incorporated into the zeolite during growth. A wide range of ion sizes can be achieved using both inorganic and organic ions.

Faujasite is an example of a zeolite. The tunnels in faujasite and in tridymite are compared in Fig. 2.8. Faujasite has a large unit cell that contains 192 silicon atoms and 384 oxygen atoms. In order to simplify the illustration and to emphasize the central point that the tunnels are much larger than those in tridymite, only the lines representing the silicon–oxygen–silicon bonds are shown. The silicon atoms are where the lines meet—the vertices. (The oxygen atoms are not shown.) The much smaller six-membered rings in tridymite are shown at the bottom. (The oxygen tetrahedron network in tridymite is shown in Fig. 2.6. The projection in Fig. 2.6 is rotated 90° with respect to the projection in Fig. 2.8.)

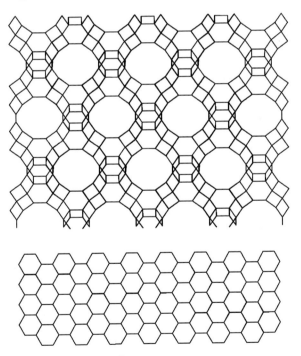

Figure 2.8 Comparison of the large tunnels in the structure of the zeolite, Faujasite, (top) and the small tunnels in Tridymite (bottom). To simplify the figure, the vertices are the positions of the silicon atoms, and the lines are the silicon–oxygen–silicon bonds (the oxygen atoms are not shown). The tunnels are formed by networks of 12 and 6 corner-sharing tetrahedrons in Faujasite and Tridymite respectively. Chemists tailor the size of the tunnels in zeolites to make catalysts and filters.

There are several dozen naturally occurring zeolites and hundreds of different synthetic zeolites. Thanks to the development of x-ray crystallography, there is now an atlas of zeolite structures documenting the wide range of pore structures that the architects of this atomic world of silicate chemistry have constructed (Baerlocher *et al.* 2001).

In the last fifty years zeolites, which started as laboratory curiosities in the eighteenth century, have become a multibillion dollar industry (Newsam 1986 and Higgins 1994). The whole field of zeolite chemistry is yet another example of how the development of science in the twentieth century is reflected in products made from sand. Let's look at a few examples of the different applications of zeolites.

Beginning in the 1930s, R. M. Barrer in London and J. Sameshima in Japan, among others, realized the potential of these materials for ion exchange and for the selective adsorption of gases and liquids. In chemistry it is important to be able to purify a material or to separate one material from another. Ion exchange is a process in which one kind of ion is preferentially adsorbed by a porous medium. In the 1950s Robert Milton at the Union Carbide Corporation made a synthetic zeolite called zeolite-A that had a much higher capacity for ion exchange than materials like charcoal or silica gel that are commonly used as adsorbents (Milton 1953).

Not only is zeolite-A a good adsorbent, but it is selective in the ions that are easily adsorbed. For example, zeolite-A adsorbs 40 times more oxygen than nitrogen at the boiling temperature of liquid nitrogen. The smaller oxygen molecule more easily enters the tunnels than does a nitrogen molecule. This difference is used to separate oxygen from air.

The drying of alcohols is another example of the selective adsorption of molecules. Water is preferentially adsorbed by zeolite-3A and alcohol is not. By passing ethyl alcohol over zeolite-3A, the water content can be lowered to 0.04 parts per million. An application of the adsorptive powers of both synthetic and natural zeolites that is closer to home is in kitty litter. The zeolite acts as a desiccant and adsorbs the water. It also adsorbs the ammonia thereby reducing the odor.

Synthetic zeolites are important supports for catalysts in the petrochemical industry. A catalyst increases the speed of a chemical reaction without being consumed in the process. Platinum is one of the most common catalysts and is used in catalytic converters in cars. However, platinum is expensive and one wants to use as little as possible. One way to do this is to incorporate the platinum in the tunnels of zeolites. To make a zeolite catalyst, platinum is exchanged for the sodium ions that are incorporated into the tunnels in the structure to achieve charge neutrality.

Catalysts are used to facilitate the chemical reactions that convert the heavier components of crude oil into lighter more volatile compounds that are used in gasoline (a process called cracking). The effectiveness of the catalyst is enhanced by making zeolites with tunnels whose sizes are matched to the molecules that are involved in the chemical reaction and, in favorable cases, controlling the orientation of the molecule with respect to the platinum inside the tunnels to further enhance the reaction. In 1962 Mobil introduced a catalyst that used a synthetic zeolite-Y, which is related to the naturally occurring mineral faujasite (Fig. 2.8). It is 1000 times better than earlier catalysts, and within a few years it became the industry standard for cracking crude oil.

Remarkably zeolites, which have much larger ratios of silicon to aluminum than zeolite-Y, act as catalysts to convert methanol to high octane gasoline. In 1997 Mobil introduced a process to make gasoline from the methane in natural gas using a high silica content zeolite, ZSM-5. The methane is converted to methanol, and the methanol is then passed over ZSM-5 at high temperatures to produce gasoline.

The exquisite control over the synthesis of zeolites can lead on the one hand to catalysts that reduce the molecular weight of the components in crude oil used to produce gasoline, and on the other hand to catalysts that can increase the molecular weight by taking a small molecule like methanol and making it into gasoline.

The next example of a material that has even larger pore sizes— 2 nm to 50 nm—is silica gel. This material is inside those ubiquitous little sacs found in pill bottles and other packing containers to act as drying agents (Fig. 2.4). Silica gel is used as a desiccant because it has an enormous surface area. A typical silica gel has a surface area that is approximately 3 million times that of a crystal of quartz.

Zeolites are crystalline materials and the sizes of the tunnels in each zeolite are uniform. However, there are limits to how large a cage or tunnel can be constructed while maintaining an ordered arrangement of cages. Silica gel can be thought of as a random network of nanoparticles that are bonded together. Each nanoparticle is composed of a random network of corner-sharing tetrahedrons. This is what leads to the enormous increase in the effective surface area of the material, because the surface area is the sum of the surface areas of all the nanoparticles.

Chemically silica gel is a porous glassy form of silicon oxide. It is synthesized by heating sand and sodium hydroxide to form sodium silicate. The sodium silicate is then dissolved in acid, and the resulting solution polymerizes to form a gel of hydrated silicon oxide. Silica gel is then obtained by heating the gel to remove the water. The process for the production of silica gel was patented in 1918 by Walter Patrick. Its initial application was in the manufacture of gas masks during the First

World War (Patrick 1918). The poison gas was preferentially adsorbed by the silica gel as the air passed over it.

In order to further increase the porosity of materials made from sand, one has to be able to prevent the gel from shrinking during drying. The shrinking is caused by the surface tension of the liquid between the nanoparticles. The surface tension pulls the particles together. The surface tension problem can be avoided if the material is dried under what are known as supercritical conditions.

At normal temperatures and pressures, a material can exist either as a liquid or as a vapor, but with increasing temperature and pressure the difference in density between the liquid and the vapor decreases. Above a temperature and pressure called the critical point, there is no difference between the liquid and the vapor and therefore no boundary and no surface tension. Drying a porous material in supercritical carbon dioxide produces materials called aerogels with effective pore sizes of greater than 50 nm (see Fig. 2.4).

Another method to make aerogels under ambient conditions was developed by Brinker and co-workers (Deshpande *et al.* 1993). The surface is coated with, for example, trimethylsilane that retains its shape during drying.

Aerogels have some of the lowest densities of any solid and therefore, the biggest pores. These novel materials find applications as extremely efficient insulators and have been used as insulation on the Mars Rover. The insulating properties are vividly illustrated in Fig. 2.9 where the heat from a torch does not melt crayons resting on the aerogel.

Another example of the use of aerogels is the Stardust mission that flew a rocket through the tail of the Comet "Wild 2." An aerogel was used to retrieve samples of the dust in the tail of the comet and return them to Earth. The rocket was traveling at an enormous speed, and the challenge was to gently slow down the dust in the tail of the comet and bring it to rest in the aerogel without melting or vaporizing the dust particles. The extremely low density of the aerogel provided the equivalent of a delicate butterfly net in that the dust particles slowly lost their kinetic energy as they passed into the aerogel and gently came to rest (Baker 2006 and Burnett 2006). The effectiveness of this butterfly net was demonstrated by the discovery of trace amounts of the amino acid glycine in the comet dust. The presence of the glycine meant that comets undoubtedly delivered pre-biotic molecules to earthlike planets from space (http://stardust.jpl.nasa.gov/news/news115.html).

In the twentieth century, materials scientists learned how to control the architecture of silicon–oxygen tetrahedrons in a wide range of porous sand-based materials, but nature has its own fascinating ways of controlling the architecture of silicon–oxygen tetrahedrons. One of the most intriguing is found in opals. Chemically, opals have a composition of SiO_2 plus varying amounts of water (1–10%). Opals have deep

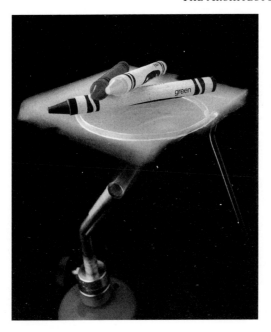

Figure 2.9 Aerogels have very low densities with pore sizes that are greater than 50 nm. They are extremely good insulators and prevent the torch from melting the crayons. (Courtesy of the Jet Propulsion Laboratory http://stardust.jpl.nasa.gov/photo/aerogel.html.)

varying colors and show spectacular flashes of color when they are rotated under a light.

The origin of this beautiful display of color was a mystery that was solved in the 1960s by J. V. Sanders and his colleagues in Australia (Sanders 1964). In the slow process of growth over the course of thousands of years, spheres of amorphous silicon dioxide are formed that are almost uniform in size. These spheres are individually composed of random networks of corner-sharing silicon–oxygen tetrahedrons. The spheres assemble into a three-dimensional close-packed array. The microscopic structure of an opal is shown in the electron microscope picture in Fig. 2.10.

It is believed that opals form in pockets in rocks. Sand and silica dissolves in water and the resulting solution fills the pockets. Over time the water evaporates, and the concentration of silica increases until the spheres separate out of the solution. If the rate of evaporation is slow enough, then the famous "fire" opals are formed, and there are large regions in the opal that have well ordered arrays of similar sized amorphous spheres.

The origin of the flashes of color in precious opals is Bragg diffraction (Box 2.1) from the ordered arrays of amorphous spheres of SiO_2.

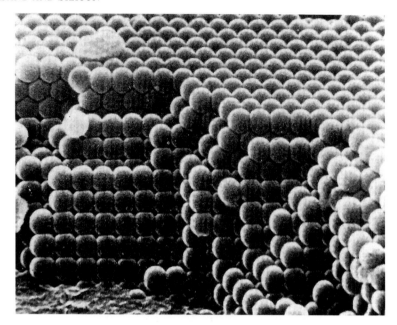

Figure 2.10 Electron micrograph showing that opals are composed of ordered arrays of amorphous spheres of silica. Bragg diffraction of light from the lattice of amorphous silica spheres leads to the flashes of light that are observed when an opal is rotated. (Photo courtesy of Hans-Ude Nissen and reproduced with the permission of the Mineralogical Society of America.)

Diffraction occurs when the wavelength of the light and the spacing between the layers of silica spheres are similar. Visible light has a range of wavelengths that varies from 380 nm (violet) to 760 nm (red), and this is comparable to the spacing of the close-packed array of amorphous spheres. Nature has conspired to grow precious opals with sphere sizes that diffract in the range of visible light so as to show colors that range from red to green.

We have been looking at the architecture of natural minerals and man-made materials originating from sand. However, the largest use of sand is in biomineralization where silica is incorporated into both unicellular algae like diatoms and protozoa-like radiolarians. Diatoms, radiolarians, and two of the three classes of sponges have skeletons composed of silica instead of the more common calcium carbonate skeletons found in animals. Industry uses tens of thousands of tons of sand per year whereas nature uses gigatons of sand in biomineralization (Perry 2003).

Diatoms are important components of the phytoplankton that are at the bottom of the food chain in oceans and rivers. They are microscopic organisms and thousands of different species have been found. Their skeletons consist of two overlapping shells called frustules. Examples are shown in Fig. 2.11.

Diatoms have a large variety of intricate and symmetric shapes. When the frustule on the right is rotated by 120°, it looks identical to the unrotated frustule. For the frustule on the left, the left and right sides are mirror images of each other. Diatoms are broadly classified according to whether they have one or the other of these symmetries.

Diatoms reproduce by the nucleus of the cell splitting, with each new nucleus moving toward one of the frustules. Then the frustules separate and a new bottom and top are grown, so each new diatom has one of the original fustules and a new one. To form a frustule, the diatom grows a sack of organic material that defines the shape of the frustule. To fill the shape, silicon that exists in the water as silicic acid is absorbed and converted to silica.

The growth of diatoms and the concentration of diatoms in rivers and marine environments are critically dependent on the concentration of silicic acid in the water. This leads to a silicon cycle of boom or bust as the concentration of silicic acid cycles from high to low throughout the year. Silicon originates in rocks that are eroded by the rivers in the spring and carried to the ocean. The diatoms then grow and use up the available silicic acid. A large portion of the diatoms are eaten by the next highest member of the food chain, and the rest die when they can no longer reproduce as a result of the depleted supply of silicic acid. In turn the dead diatoms decay. The remaining silica skeletons either dissolve, supplying new silicic acid, or they sink to the bottom (Yool and Tyrrell

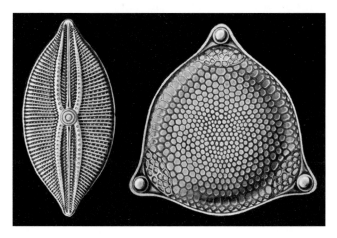

Figure 2.11 Examples of the exquisite geometry of the silica skeletons of diatoms that range in size from 2 to 200 micrometers. From the drawings of Ernst Haeckel which appeared in *Kunstformen der Natur*. Bibliographisches Inst., Leipzig (1904).

2003). The production of silicic acid is intimately related to the sequestration of carbon dioxide. Carbonates dissolved in water react with sand producing silicic acid and CO_2 that is subducted into the Earth's mantle and released millions of years later in volcanoes.

Over time the skeletons have collected on the bottom of rivers and oceans to form what we call *diatom*aceous earth. It is used as a filter material and as an adsorbent. It is often spread around to clean up after chemical spills, and it is used in some types of kitty litter.

Another use of diatomaceous earth brings us back in an unusual way to science in the twentieth century. The most important recognition of outstanding science is the Nobel Prize, which was first awarded at the beginning of the twentieth century in 1901. The prize was endowed by Alfred Nobel who made his fortune from his invention of dynamite. In the mid 1800s the explosive nitroglycerine had been invented, but it was very unstable. Nobel had the idea to absorb the nitroglycerine in diatomaceous earth and make it into sticks of what we now call dynamite (Nobel 1868). So the Nobel Prize can trace its origin to the incorporation of silica—sand—in the skeletal remains of diatoms into dynamite.

Our brief tour of the structures of the different materials in Fig. 2.4 shows how the architecture of corner-sharing silicon–oxygen tetrahedrons varies from rings to helices to ordered or random networks with ever increasing pore sizes. Nature, on the other hand, makes materials like opals with their own unique type of order and incorporates silica in the skeletons of aquatic microorganisms. The ability of scientists to control the architecture of sand is in part based on being able to determine the structures of these materials using x-ray diffraction.

The challenge for materials scientists in the twenty-first century is to understand how nature manages to grow a wide variety of intricate structures out of silica. That knowledge could enable new man-made structures to be fabricated to meet new technological demands.

3
How Pure is Pure?

Our language is full of expressions of purity such as "pure as the driven snow." Diamonds are admired because they are seen as pure and flawless, but single crystals of silicon are the clear winner in terms of purity and perfection. The electronics industry takes sand, usually in the form of quartzite, and transforms it into large single crystals of silicon that are the most chemically pure and structurally perfect crystals on Earth. It is fair to say that the whole electronics industry, as we know it, would not exist if scientists hadn't developed processes to make pure silicon from sand. This chapter discusses how sand becomes ultra-pure silicon and the science behind making pure silicon.

For decades Ivory soap advertised that it was 99.44% pure and that it was "safe for your baby's bottom"; 99.44% pure means that if it were possible to count out 10,000 soap molecules, then 56 of them would be impurities of some kind. In contrast to Ivory soap, silicon is routinely purified to a level of a part per billion or 99.999999% pure. To appreciate that number, consider a small bucket used by children at the beach. The bucket contains about a billion grains of sand. Silicon made with a purity of a part per billion is equivalent to one grain in the bucket.

The electrical properties of silicon are particularly sensitive to even small amounts of impurities, and the transistors in computer chips depend on the controlled addition of impurities to silicon. To make reliable and reproducible computer chips, the purity of the starting silicon had to be much higher than had ever been achieved.

After the invention of the transistor in 1947 at Bell Telephone Laboratories Inc., two processes were developed. The first produced laboratory scale quantities for use by the research and development communities, and the second led to the current industrial scale production of the thirty odd tons of silicon used by the electronics and solar cell industries each year. The first technique is zone refining, invented by Bill Pfann at Bell Labs (Pfann 1951). The second involves high performance fractional distillation, developed by Siemens (Wilkes 1996). Both techniques revolutionized the ability of industry to make ultra-pure materials. Today we take for granted that ultra-pure materials will be readily available, but 50 years ago achieving purities of parts per billion represented a revolutionary development in materials science.

The underlying concept in the purification of materials by zone refining is that there is a difference between the concentration of impurities in the liquid phase and the solid phase of a material when they coexist at the melting temperature. Similarly, in fractional distillation the concentration of impurities in the liquid and vapor phases will be different when they coexist at the boiling temperature.

Consider an everyday substance—salt water. The ocean salt water contains about 3.5% salt. The presence of salt in seawater lowers its freezing temperature to between 0 °C and –8.7 °C depending on the concentration of the salt. This leads to all sorts of useful applications, from salting roads in winter to making ice cream in summer. When salt, water, and ice are mixed together, some of the ice melts. The temperature goes down because heat is drawn from the salt water to melt the ice. On the one hand, melting the ice on the road makes driving safer. On the other hand, lowering the temperature of the salt ice mixture in the ice cream maker causes the cream to freeze. When sea water freezes, the resulting ice contains much less salt than the salt water. The ice rejects the salt on freezing if the temperature is between –22 °C and 0 °C.

Next consider the boiling of seawater. When it is boiled, most of the salt is left behind and the condensed steam is purer than the original seawater. In fractional distillation this process is repeated again and again with each successive condensate being purer than the previous one. Pure drinking water can be obtained this way. In the Siemens process to produce ultra-pure silicon from sand, an impure silicon-containing compound, trichlorosilane ($SiHCl_3$), is distilled. Each successive condensate has a lower concentration of impurities. Finally, pure trichlorosilane can be converted to ultra-pure silicon.

The lowering of the freezing temperature of salt water and the rejection of the salt in the formation of ice are manifestations of the same thermodynamic principles that lead to the purification of silicon by zone refining. In zone refining a solid is heated and cooled so that it melts and freezes. In silicon containing a small amount of an impurity, the concentration of the impurity will be different in the solid phase than it is in the liquid phase when they coexist. This difference can be exploited to remove the impurities.

Zone refining was developed by Bill Pfann to purify the germanium (another element with properties similar to silicon) that was used in early transistors. In his early apparatus, he placed a rod of germanium in a pure graphite boat that was slipped inside a long quartz tube. To melt a narrow cross-section or zone in the germanium, he put a radio frequency coil around the tube. An alternating current passing through the coil melts the germanium in the zone by induction. The familiar microwave oven heats by the same principle except that it operates at much higher frequencies. The coil is slowly moved from one end of the graphite boat to the other. As the zone moves down the rod, germanium at the leading edge of the zone melts and germanium at the trailing edge of the zone freezes. The germanium that solidifies at the trailing edge of the zone has a lower impurity concentration. As the zone of liquid germanium moves down the rod, the concentration of impurities in the liquid increases, and the impurities are essentially pushed down the rod of germanium.

Bill Pfann realized that the process would be much more efficient if he used more than one coil. In his first apparatus he had six coils moving down the germanium. As will be discussed in Chapter 4, the electrical conductivity of semiconductors is very sensitive to the presence of trace impurities. George Dacey measured the conductivity of various parts of the germanium sample at low temperatures and concluded that there was less than one impurity atom per 10 billion germanium atoms (10^{10}) at the purified end of the silicon rod (Millman 1983, p. 423).

The purification of silicon required a bit more finesse. Silicon melts at 1420 °C and, at these high temperatures, the silicon being purified by zone refining would be contaminated by impurities that leach out of the container holding the silicon. The problem is to melt and freeze the silicon without it contacting a container. Henry Theuerer at Bell Labs solved the contamination problem when he invented what is called the float zone process (Theuerer 1952). He had the ingenious idea of holding a rod of impure silicon vertically, as shown in Fig. 3.1. A heater melts the zone of silicon, and the surface tension of the liquid silicon is strong

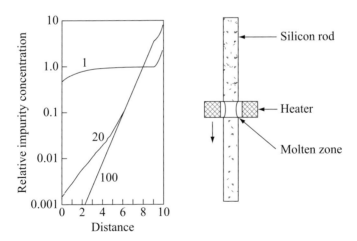

Figure 3.1 Zone refining is used to purify silicon. The heater melts a thin zone in a rod of impure silicon, and the liquid zone moves as the heater is translated along the rod. The liquid phase has more of the impurity than the solid phase so the silicon that solidifies at the trailing edge of the zone has fewer impurities. The impurities are pushed along the rod as the liquid zone moves along the rod. The impurity concentration along the rod is shown on the left after 1, 20, and 100 passes of the zone. (Adapted from Pfann 1966.)

enough to prevent the liquid from leaking out. The silicon rod is clamped at both ends, but it is not in contact with a container in the middle. The float zone process yielded silicon of comparable purity to germanium.

Figure 3.1 illustrates the concentration of the impurity along the rod after one pass of the heater. The concentration at the top is substantially lower, and the concentration at the bottom is substantially higher. With repeated passes, the purity at the top of the rod increases, and the improvement in purity extends further and further along the rod. In essence the impurities are swept along the rod by repeated passes of the heater. The curves in the figure show the impurity distribution along the rod after 1 pass, 20 passes, and 100 passes.

The final purity depends on the ratio of the impurity concentration in the solid silicon divided by the concentration in the liquid silicon and on the speed at which the zone moves along the rod. The curves in the figure are calculated for a ratio of 0.1. The concentration of the impurity at the beginning of the rod can be reduced by 15 orders of magnitude (10^{-15}) (Pfann 1966).

A flow chart for the industrial scale production of silicon by the Siemens process is illustrated in Fig. 3.2. The process starts with metallurgical grade silicon because it is the least expensive. The raw material for the production of metallurgical grade silicon is sand.

The process involves the production, purification, and then decomposition of trichlorosilane. This compound of silicon is chosen because it has a low boiling temperature of 31.8 °C and therefore can be more easily purified by fractional distillation. The process produces rods of

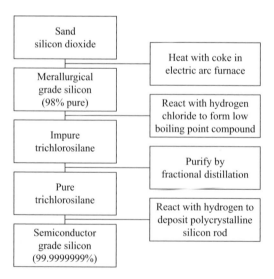

Figure 3.2 Flowchart for the conversion of sand into semiconductor grade silicon using the trichlorosilane process developed by Siemens.

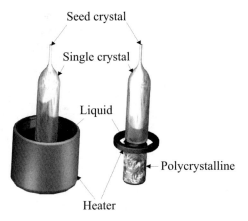

Figure 3.3 Czochralski (left) and float zone (right) growth of single crystals of silicon. In Czochralski growth, a seed crystal is slowly withdrawn from a crucible containing liquid silicon. In float zone growth, a heater melts a narrow zone between a polycrystalline rod of silicon and a single crystal seed rod. As the liquid zone is slowly lowered, a single crystal rod of silicon is grown.

ultra-pure polycrystalline silicon, and it is the leading process for the production of semiconductor grade silicon (Wilkes 1996).

The trichlorosilane process produces pure silicon in the form of a rod composed of microcrystallites. Finally, the rod is converted into a large perfect single crystal from which slices (wafers) are cut for the production of semiconductor devices. Large single crystals of silicon are grown by the Czochralski technique and the float zone techniques that are illustrated in Fig. 3.3.

In the Czochralski technique, pure silicon is melted in a quartz boat and a seed crystal is touched to the surface. The seed crystal provides a template on which more silicon crystallizes as the seed is slowly withdrawn. A large single crystal of silicon grown by the Czochralski technique is shown in Fig. 3.4.

The float zone technique is an outgrowth of the zone refining purification process shown in Fig. 3.1 in which a rod of semiconductor grade polycrystalline silicon is held in a vertical position. A single crystal seed rod is aligned above the polycrystalline rod, and a heater melts a narrow zone at the top of the polycrystalline rod. The molten silicon crystallizes on the seed rod and, as the zone moves down, the polycrystalline rod is converted into a single crystal.

Silicon is unique in forming extremely perfect crystals. The structure of a crystal is described in terms of building blocks or unit cells that are assembled to form a macroscopic crystal. However, most materials crystallize with defects or imperfections in this idealized structure. Some defects occur in the form of extra or missing atoms. Others, known as dislocations, involve extra or missing rows of atoms. As a

Figure 3.4 A single crystal boule of silicon grown by the Czochralski technique. The silicon crystal grows from the seed crystal at the top, and it is suspended from the seed during growth. The perfect crystal of silicon is taller than the technician who is measuring the temperature of the crystal. (© 2000 Kay Chernush.)

result, adjacent regions of the crystal are slightly out of position with respect to each other.

Both the Czochralski and float zone techniques can, with great care, produce almost defect-free single crystals. Producing thousands of tons of perfect crystals of silicon every year is a testimony to the skill of twentieth century engineering. Each of these crystals weighs more than 60 kg (130 lbs), has a purity of 99.9999999%, and has virtually no defects.

The electronics industry and the information age rest on the foundation of ultra-pure, defect-free silicon crystals made from sand. In turn, the production of ultra-pure silicon rests on the foundation of the laws of thermodynamics that were developed during the nineteenth century. The evolution of these laws is outlined in Box 3.1. The rest of this chapter is devoted to the thermodynamic principles behind fractional distillation that is used to purify silicon.

To explore these principles, consider two everyday substances—water and acetone. Depending on the temperature and the pressure, each of these materials can exist either as a solid, a liquid, or a gas. These are referred to as different phases of a material. A phase is a homogeneous and distinct part of a system that is separated from other parts of the system by definite bounding surfaces. Melting is the phase transition from the solid phase to the liquid phase, and boiling is the phase transition from the liquid phase to the gas or vapor phase. The laws of thermodynamics govern how these different types of phase transitions depend on the temperature, pressure, and composition.

How does the temperature at which a phase transition occurs change as the pressure changes? At sea level, water and acetone boil at 100 °C and 56.5 °C respectively. At what temperature will each of them boil at the top of Mount Everest? Once we understand the relation between the boiling temperature and pressure of water and of acetone, then we can ask what happens when we have a mixture of water and acetone. This in turn reveals the principle behind fractional distillation that can be used to separate acetone from water or to purify trichlorosilane.

Three cylinders are shown in Fig. 3.5, and they illustrate the transition from the liquid phase to the vapor phase as a function of pressure at constant temperature. On the left a liquid fills a cylinder that is closed by a movable piston. The piston applies pressure compressing the liquid. As the pressure on the piston is decreased, a pressure that is called the vapor pressure is reached where the liquid and vapor phases coexist as shown in the middle cylinder. The volume increases as the liquid transforms into the vapor phase because the volume per unit mass of the

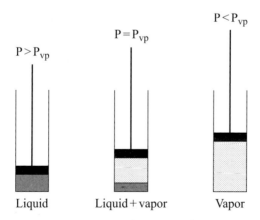

Figure 3.5 A transition from the liquid phase to the vapor phase occurs when the pressure decreases as the piston is withdrawn. At a given temperature there is a critical pressure, called the vapor pressure, at which the liquid and vapor phases coexist.

vapor is much larger than that of the liquid. As a result, the piston continues to move up in the cylinder under an applied pressure equal to the vapor pressure until all of the liquid has transformed into vapor. Decreasing the pressure further causes the vapor (gas) to expand, lowering its density (right cylinder).

The process can be reversed and the pressure increased, driving the piston into the cylinder and compressing the gas. Again, when the external pressure equals the vapor pressure, the vapor phase transforms into the liquid phase, and the piston falls until all the gas (vapor) has transformed into the liquid. Further application of pressure then compresses the liquid.

Thermodynamics is about the relation between work and heat. In the example above, mechanical work is done on the system as the piston is either pushed or pulled. For the liquid to boil or the vapor to condense, heat has to be added or removed to keep the temperature constant. The first law of thermodynamics (Box 3.1) requires that the mechanical work done by moving the piston is equal to the heat transferred to the liquid. In each case mechanical work is transformed into heat.

The second law of thermodynamics deals with the opposite transformation—using heat to do mechanical work. In the previous example, the system of water, cylinder, and surroundings (usually called the reservoir) were held at a constant temperature, and mechanical work was done by moving the piston leading to a transfer of heat.

To use heat to do mechanical work, heat has to transfer from the reservoir to the water. If both are at the same temperature, there is no reason for heat to spontaneously transfer from the reservoir because this would cause the temperature of the water to rise. It is like water flowing in a river—water flows downhill, it doesn't flow uphill. There has to be a driving force in the form of a temperature difference between the reservoir and the water. Heat flows from the reservoir at a high temperature to the water at a lower temperature. This causes the water to boil and, in turn, do work.

In 1824 Sadi Carnot proved that in this process not all of the heat leads to useful work. He was interested in determining the maximum efficiency of a steam engine. Efficiency is the ratio of the work produced by the engine divided by the amount of heat that is supplied. In a steam engine, water at one temperature is heated to a second higher temperature causing the water to boil. The resulting steam does mechanical work by pushing a piston. The steam is then condensed and returns to the first temperature.

Carnot envisaged a hypothetical ideal heat engine in which each cycle of the engine was carried out by only taking small steps that could go forward or backward while keeping the system at equilibrium. This is called a reversible engine because the cycle could be run in either direction. In the forward direction it acts as a heat engine, and in the reverse direction it is a heat pump.

Carnot proved that the efficiency of his ideal, reversible heat engine only depended on the temperature difference in the different legs of the cycle divided by the absolute temperature of the steam (the absolute temperature scale is discussed in Box 3.1). The efficiency was independent of the material used in the engine. The idealized heat engine had the highest possible efficiency and therefore any real steam engine would have various inefficiencies and losses that would lead to a lower efficiency.

The conclusion drawn from the ideal heat engine is one way of stating the second law of thermodynamics. The law states that a process that just takes heat from a reservoir and converts it to work without a change in temperature is impossible. Furthermore, a heat engine that takes heat at one temperature and delivers it at a lower temperature cannot do more work than the ideal heat engine. Building on the work of Sadi Carnot, scientists in the middle of the nineteenth century developed the mathematical framework of thermodynamics that related changes in the energy of a system to changes resulting from the addition of heat to the system or from work done on the system.

Returning to the examples of water and acetone, if the temperature is changed, then how will the vapor pressure change? For example, if we know the boiling temperature of water and acetone at sea level, can we predict the boiling temperature at the top of Mount Everest? The second law of thermodynamics establishes a relation between the vapor pressure, the boiling temperature, and the heat of vaporization of water and acetone. The mathematical relation, the Clausius–Clapeyron equation, is derived in Box 3.1.

The vapor pressure of water and acetone and the temperature is shown in Fig. 3.6. (The vapor pressure varies by two orders of magnitude in this temperature range, so for clarity the logarithm of the vapor pressure is plotted against the temperature.) The two curves were calculated using the Clausius–Clapeyron equation and the measured heat of vaporization of each liquid and their boiling temperatures at atmospheric pressure. The vapor pressure of acetone is higher than that of water at all temperatures. The two liquids boil at 100 °C (373 °K) and 56.5 °C (329.7 °K) where the vapor pressure of water and acetone equal atmospheric pressure at sea level. On the other hand, atmospheric pressure on the top of Mount Everest is approximately one third of atmospheric pressure at sea level. This means, according to Fig. 3.6, that the vapor pressure of water and acetone will equal atmospheric pressure and boil at the top of Mount Everest at around 70 °C (343 °K) and around 25 °C (298 °K) respectively.

The curves in Fig. 3.6 demonstrate a more fundamental thermodynamic result about phase transitions. Each shows that if a system is composed of a single chemical substance and has two phases coexisting

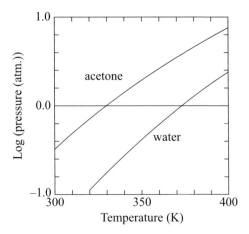

Figure 3.6 Vapor pressure of acetone and water versus temperature in degrees Kelvin. The pressure is given on a logarithmic scale and it runs from 0.1 atmospheres to 10 atmospheres. At atmospheric pressure (Log(1)=0) acetone and water boil at 329.6 °K (56.5 °C) and 373.1 °K (100 °C) respectively. At temperatures in between, the vapor pressure of acetone is greater than 1 atm and the vapor pressure of water is less than 1 atm.

in equilibrium (liquid and vapor), then it has only one degree of freedom. Once one variable, the vapor pressure or the temperature, is determined, then the other variable is fixed by thermodynamics.

This is an application of a fundamental rule of thermodynamics called the "phase rule" that conveniently summarizes the conditions for equilibrium in a system with multiple components and phases. The phase rule limits the number of variables that have to be specified in order to describe the purification of a water/acetone mixture or to purify silicon via fractional distillation.

The phase rule was developed by J. Willard Gibbs to determine the number of variables that have to be specified in order to completely define equilibrium in a chemical system. He called the number of necessary variables the degree of freedom (F). A chemical system can be composed of a number of components (C)—in our example there are two: water and acetone. They can be distributed among a number of different phases (P). In our example of the boiling of a mixture of water and acetone there are two distinct phases: liquid and vapor. The Gibbs Phase Rule states that the number of degrees of freedom is equal to the number of components minus the number of phases plus two, that is $F = C - P + 2$. So for a mixture of water and acetone that is boiling, the number of degrees of freedom is two ($F = 2 - 2 + 2$). Once we have specified the temperature and average composition of the mixture, then the fraction of each component that is in the liquid or the vapor phase is fixed by thermodynamics if the system is in chemical equilibrium. The relation between the components and the phases is expressed as a graph called a phase diagram. It is the basis for the purification processes described earlier in the chapter.

It is worth taking a small detour to talk about Josiah Willard Gibbs because he is one of the greatest American physicists. He published a series of seminal papers on thermodynamics between 1876 and 1878 entitled "On the equilibrium of heterogeneous substances" (Gibbs 1876). The articles deal with almost every conceivable thermodynamic property of a system. Gibbs presented the fundamental thermodynamics of gases, mixtures, surfaces, solids, phase changes, chemical reactions, electrochemical cells, sedimentation, and osmosis. Gibbs went on to be one of the founders of the field of statistical mechanics with the publication of his book *Elementary Principles in Statistical Mechanics* (Gibbs 1902). Finally, every student of physics or engineering has learned the techniques of vector analysis that are based on *Elements of Vector Analysis* that was published privately by Gibbs for his students between 1881 and 1884 and subsequently was turned into a textbook by one of his students (Wilson 1901). This is a tremendous body of original research emanating from one genius working essentially alone throughout his life.

Gibbs was born in New Haven Connecticut on 11 February 1839. His father, who had the same name, Josiah Willard Gibbs, was a professor of sacred languages in the Divinity School at Yale University. His son graduated from Yale and earned the first Ph.D. in engineering in the United States in 1858. As with many American scientists at that time, he spent three years in Europe studying at universities in Paris, Berlin, and Heidelberg before returning to become a professor of mathematical physics at Yale.

He stayed at Yale throughout his whole career, but not without an interesting incident. Gibbs was supporting himself from a small inheritance and was completely immersed in his research. Yale did not even give him a salary, even though he was a professor of mathematical physics. In 1879 he was invited to give a series of lectures at Johns Hopkins University. The university was in the process of trying to expand the physics department, and they immediately offered Gibbs a professorship with a good salary. Word trickled back to Yale that they were about to lose their most outstanding physics professor. Eventually Yale stepped up to the plate. In 1880 they offered Gibbs a salary of $2000 per year and convinced him not to leave.

Gibbs' contributions to science have been far reaching, and yet he was relatively unknown outside of the physics community because he chose to work almost entirely on his own. He was born in New Haven, lived and worked in New Haven, and died in New Haven on 28 April 1903. His world was his research, and he hardly strayed from the physics department and the home of his sister and her family with whom he lived a few blocks away. He was completely focused on his research, and it sustained him for over 45 years.

Now let us turn to a mixture of water and acetone and ask: how does a mixture boil? Although the thermodynamics are completely described in Gibbs' treatise, let's begin with experiments done at almost the same time on the boiling of salt water. In 1878 François-Marie Raoult, a professor of chemistry at the University of Grenoble in France, demonstrated that the vapor pressure above a solution containing water and various salts decreases in proportion to the fraction of the liquid that is water as opposed to salt.

The lowering of the vapor pressure over the solution has a simple physical interpretation. The pressure over the liquid is a function of the average tendency of molecules to escape from the liquid into the vapor multiplied by the number of water molecules. In a solution some of the water molecules have been replaced by a second component and therefore there are fewer water molecules to escape. As a result, the vapor pressure of water over the solution is reduced in proportion to the fraction of water molecules in the solution. At the same time the vapor pressure of salt is very low and contributes very little to the total vapor pressure.

The boiling temperature of water decreases on going from sea level to the top of Mount Everest because the pressure of the atmosphere decreases and, as a result, the vapor pressure of water reaches atmospheric pressure at the top of Mount Everest at a lower temperature. However, the boiling temperature of water at constant pressure increases with the addition of salt because the vapor pressure has decreased, and the temperature has to increase to bring the vapor pressure above the liquid up to atmospheric pressure.

Unlike salt water, the vapor pressures of water and acetone are not too different (Fig. 3.6). When water and acetone are mixed together to form a solution, let us assume that they do not interact with each other. Chemists refer to this as an ideal solution. In this case, the fraction of the mixture that is water will act like pure water and, similarly, the fraction of acetone will act as pure acetone. Consistent with Raoult's observations for salt solutions, the part of the total vapor pressure—the partial pressure—coming from the water will be the vapor pressure of pure water multiplied by the fraction of the molecules in the mixture that are water, and similarly for the partial pressure of acetone. A number of solutions follow this simplified picture of a liquid. In those solutions that do not, they do approach the ideal behavior at the low concentrations that are relevant for the purification of silicon.

For the mixture of water and acetone to boil, the total vapor pressure must equal atmospheric pressure. The total pressure is the sum of the partial pressure contributed by each component in the liquid. The vapor pressure of acetone is higher than that of water, so acetone will contribute a larger fraction of the molecules to the vapor phase. At the same time, the number of molecules in the overall system is fixed. To reach equilibrium at the boiling temperature, the fraction of each component that is in the vapor and liquid phases will change. The net result is that the vapor is richer in the component that has a higher vapor pressure and smaller in the liquid than expected based on the overall composition.

The composition calculated for each phase as a function of temperature at atmospheric pressure is presented in Fig. 3.7. The composition of the vapor and the composition of the liquid at each temperature are given as a function of the overall composition of the mixture. The two curves represent the composition of the liquid (bottom curve) and the vapor (top curve) as the temperature is varied from the boiling temperature of pure acetone to the boiling temperature of pure water. For example, as shown in the phase diagram, in a solution that is 50% water and is at a temperature of 350 °K, the fraction of water in the vapor phase is 25% and the fraction of water in the liquid phase is 62%. The composition of the liquid is richer in water than the original solution and the composition of the vapor is poorer in water.

If a portion of the vapor is separated and condensed, then it can be used as the starting material for further purification. If that fraction is

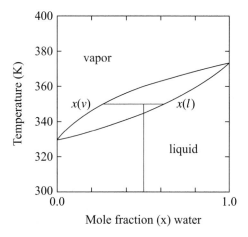

Figure 3.7 Phase diagram of a solution of acetone and water showing the mole fraction of water in the vapor phase and the liquid phase as a function of temperature. (The mole fraction is the fraction of the molecules in the solution that are water as opposed to acetone.) If a solution that is 50% water is heated to 350 °K, then the vapor and liquid phases coexist with the vapor being 25% water and the liquid being 62% water.

heated to 340 °K, then the resulting vapor will be only 10% water and 90% acetone. This process can be repeated as many times as needed to obtain the desired purity of acetone. Conversely, successive liquid fractions can be separated to obtain pure water.

In fractional distillation the container is open rather than closed and the vapor is allowed to escape. The temperature increases as the composition of the liquid phase shifts to the right in the figure. This corresponds to sliding up the liquid curve in Fig. 3.7 as the distillation progresses. A column is designed with a number of baffles that have progressively lower temperatures on going up the column. As a result, vapor will condense on the baffle and the resulting liquid, which is richer in water, will flow down the column. The net result is that the vapor gets richer in acetone as it moves up the column, and the liquid at the bottom of the column becomes richer in water.

To purify silicon the compound trichlorosilane is made. It will contain small amounts of impurities and, just like water in acetone, the impurity phases will have different vapor pressures at any given temperature. The concentration of impurities will be very small, meaning that the process will start at compositions very far to the left in an equivalent phase diagram. The principle remains the same that the impurities are left behind during distillation if the impurities have a lower vapor pressure than trichlorosilane.

A similar situation occurs in the freezing and melting of a mixture of materials, and the resulting phase diagram looks very much like the one shown above for boiling. In this case, the solid phase is stable at low temperatures and the liquid phase at high temperatures. Similar differences between the composition of the solid solution and the liquid solution are observed if the system is in equilibrium. Again, purification of one or the other component can be achieved by the same process, and by analogy it is called fractional crystallization. Zone refining consists of repeated steps of fractional crystallization.

In this chapter, we have shown how sand can be converted to the ultra-pure silicon needed by the electronics industry. The process of purification by zone refining or fractional distillation grew out of the application of the laws of thermodynamics. These laws evolved over several centuries starting with the study of the compressibility of gases by Robert Boyle in the seventeenth century and progressing to the study of the equilibrium of heterogeneous substances by J. Willard Gibbs at the end of the nineteenth century (Gibbs 1961). The laws of thermodynamics are the foundation on which twentieth century materials science and chemistry rest, and they are reflected in the purification of silicon made from sand.

BOX 3.1 The Second Law of Thermodynamics

In the historical development of thermodynamics, the laws governing the properties of gases were studied in the seventeenth and eighteenth centuries. Then the relations between different phases (solid, liquid, and gas) of a material were elucidated in the first half of the nineteenth century leading to the first and second laws of thermodynamics. Finally, the relationships between different phases of mixtures of materials were formulated at the end of the nineteenth century. They laid the foundations of twentieth century chemistry and metallurgy and for the purification of silicon.

Our understanding of the properties of gases started with observations of the relations between the pressure, volume, and temperature of gases in the seventeenth century. Robert Boyle, a British scientist who is considered by many to be the founder of modern chemistry, reported his studies on the compressibility of air in 1662. He found that the product of the pressure multiplied by the volume of a gas is equal to a constant independent of the gas. Guillaume Amontons, a French scientist and instrument maker, studied the relation between the volume and the temperature of gases in 1702. Based on these measurements he speculated that if the temperature could be reduced sufficiently then the volume would go to zero. This is the first suggestion that there is an absolute zero temperature. Almost a century later, Jacques Charles and independently Joseph Louis Gay-Lussac made accurate measurements of the change in the volume of a gas with changes in temperature. They found that the volume of a gas is directly proportional to the temperature if the temperature is given on an absolute scale ($T = 0°$ Kelvin $= -273.16$ °Centigrade). Their early measurements gave an absolute temperature of -273 °C that is very close to the presently accepted value of -273.16 °C. In 1811, Amedeo Avogadro, an Italian scientist at the University of Turin, proposed that equal volumes of ideal gases contain equal numbers of molecules. When combined these results established the ideal gas law. This law relates the pressure (P), volume (V), quantity (N), and temperature (T in degrees Kelvin) of a gas via one simple equation with a universal constant (R). The product of the pressure of a gas multiplied by its volume equals a universal constant multiplied by the product of the quantity of gas multiplied by the temperature ($PV = NRT$).

In the middle of the nineteenth century, James Clerk Maxwell and Ludwig Boltzmann among others developed the kinetic theory of gases that gives a microscopic description of the ideal gas law. The theory is based on elastic collisions of small particles, in which they exchange kinetic energy with one another but there is no change in the total kinetic energy of the particles. The collisions of millions and millions of particles against the walls of the container produce the pressure. If the volume is increased, then the pressure decreases because the particles on average hit the container walls less often.

Temperature, on the other hand, is a measure of the average kinetic energy of the particles. The kinetic energy is proportional to the square of the velocity of the particles. Absolute zero is the temperature at which the average kinetic energy of the particles in a gas (and therefore their velocities) approach zero, and if the gas didn't change to a liquid or a solid, then the volume would equal zero.

(cont.)

BOX 3.1 Continued

The kinetic theory of gases is a grand achievement of classical mechanics in the middle of the nineteenth century. It describes many different properties of gases in terms of the straightforward mechanics of the motion of a collection of particles. It is like a three-dimensional game of pool with the pool balls bouncing off each other and off the cushions (walls) that confine the balls.

One of the most important gases from the industrial point of view is steam. In a steam engine, heat is supplied to the boiler to produce steam at high pressure and then the steam drives the pistons. Heat is extracted to condense the steam to liquid water, and water is restored to high pressure via a pump. The net effect is that heat drops from high temperature to low temperature with work output to the pistons and smaller work input to the pump. With the development of steam engines, there was a growing interest in the relation between heat and mechanical work and in maximizing the efficiency of steam engines. In 1824 Sadi Carnot determined the maximum efficiency of an idealized steam engine.

In the early nineteenth century scientists did not understand that heat is another form of energy. They believed that heat was a weightless fluid called a "caloric" that flowed downhill like water. Carnot believed in the caloric theory, and his thinking about the efficiency of steam engines was influenced by work that his father, also an engineer, had done on the efficiency of water wheels. His Father, Lazare Carnot, published a memoir in 1783 on the mechanics of waterwheels. His main idea was that accelerations or shocks to the moving parts of machinery are to be avoided because they lead to various mechanical losses like friction. In an ideal machine, power is transmitted continuously in small steps, and Sadi Carnot set out to devise an idealized heat engine that operated in a similar manner. He also made the analogy with a water wheel in which the greater the fall of water, the greater the work output per unit of water input (Cropper 2001).

To visualize the Carnot cycle, let us return to the boiling of a liquid by putting Fig. 3.5 into graphical form (Fig. 3.8). The figure shows the continuous variation of the pressure and volume at two different temperatures with a difference of ΔT. On the left, as the liquid phase is compressed the volume decreases. On the right, the vapor phase expands as the pressure decreases. In between the pressure remains constant, and the volume changes as the liquid transforms to the vapor.

In the inset, the lines of the plateaus for each temperature are connected to form a parallelogram. The parallelogram represents a cycle that is superficially similar to the cycle of a steam engine. Starting at point 1 the vapor expands, drawing heat from a heat reservoir. This is called isothermal expansion because it occurs at constant temperature. Next the heat reservoir is removed and the gas is expanded, going from point 2 to point 3. The expansion causes the vapor to cool by $\Delta T°$. This is called an adiabatic expansion because no heat is exchanged with the surroundings. Next the vapor is compressed as it transforms back into the liquid phase while giving heat back to a reservoir at a lower temperature (point 3 to point 4). Finally, the system is returned to the starting temperature by adiabatic compression of the liquid.

Carnot devised a heat engine with a cycle similar to the one in Fig. 3.8 except that the different steps in the cycle were all carried out in the gas phase. His cycle used an ideal gas where the amount of mechanical work and the change in heat for each leg of the cycle can be calculated using the ideal gas law

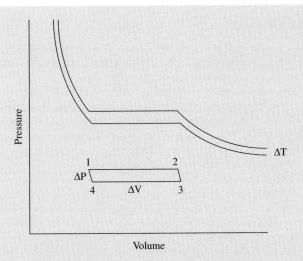

Figure 3.8 Graph of the change in pressure as a function of volume at two temperatures that differ by ΔT. The parallelogram in the inset shows a cycle in which liquid is transformed into vapor followed by a change in temperature (1 → 2 → 3). The vapor is transformed back into a liquid and the temperature changed back to the starting temperature (3 → 4 → 1). The work done in the cycle is the product of ΔV × ΔP. As discussed in the text, this product is equal to the heat added to the liquid in the first leg (1 → 2) times the difference in temperature, ΔT divided by the temperature, T. This equality is a statement of the second law of thermodynamics that was used to calculate the vapor pressure in Fig. 3.6.

(Feynman et al. 1975). The calculation showed that the ratio of the heat added in leg one, Q_1, divided by the temperature, T_1, is equal to the ratio of the corresponding quantities in leg three, that is $Q_1/T_1 = Q_2/T_2$.

According to the first law of thermodynamics, the net mechanical work equals the net heat supplied to the system. For the Carnot cycle, this is the heat put into the gas in the first leg of the cycle minus the heat taken out of the system in the third leg, that is $(Q_1 - Q_2)$. (Carnot's analysis did not rely on the first law of thermodynamics because it was only developed years later, but the analysis is simplified by using the first law.)

The efficiency of a heat engine is the net heat supplied while going around the cycle $(Q_1 - Q_2)$ divided by the heat put into the system in the first leg of the cycle (Q_1). Using the result above that $Q_1/T_1 = Q_2/T_2$, the efficiency equals $(Q_1 - Q_2)/Q_1 = (T_1 - T_2)/T_1$. In calculating the efficiency of a heat engine based on an ideal gas, Carnot had made a fundamental new discovery. Even the most efficient heat engine cannot do work if there is no change in temperature. His result also means that all heat engines have an efficiency that is less than one.

The conclusions of the Carnot cycle can be used to derive the relation between the change in the vapor pressure when the temperature is changed. The parallelogram in the inset in Fig. 3.8 is approximately a Carnot cycle. The result

(cont.)

BOX 3.1 *Continued*

above for a Carnot cycle is that the net heat supplied to the system is $Q_1(\Delta T)/T_1$. The heat needed to boil one mole of liquid is the heat of vaporization and is given the symbol ΔH_v, so the net work is given by $\Delta H_v(\Delta T/T_1)$.

Mechanical work is given by the integral of the pressure multiplied by the change in volume. Graphically, it is the area under the curve for each leg of the cycle in Fig. 3.8. In the first and second legs, the work is positive because the volume is increasing. In the third and fourth legs, the work is negative because the volume is decreasing. The net work is the area of the parallelogram formed when the area under the curves of legs 3 and 4 are subtracted from the sum of the areas under the curves for legs 1 and 2. So the net work is graphically the product of the change in volume, ΔV, and the change in pressure, ΔP.

To satisfy the first law of thermodynamics, the net thermal work must equal the mechanical work, and this gives the relation that the product of the change in volume multiplied by the change in pressure is equal to the heat of vaporization multiplied by the fractional change in temperature ($\Delta V \Delta P = \Delta H_v \Delta T/T$). This relation between changes in pressure, volume, and temperature is of fundamental importance. It is an alternate statement of the second law of thermodynamics, and most of thermodynamics can be derived from this equation and the first law of thermodynamics (Feynman *et al.* 1975).

When this relation is rearranged as the change in the vapor pressure divided by the change in temperature, it is called the Clausius–Clapeyron equation. It was used to calculate the vapor pressure curves in Fig. 3.6 using the ideal gas law, the measured boiling temperatures at one atmosphere, and the heats of vaporization for acetone and water.

The development of thermodynamics sketched above is historical in nature, and a modern description, which comes out of the work of Ludwig Boltzman, J. Willard Gibbs, and others, emphasizes the statistical nature of thermodynamics. The quantity Q/T in the Carnot cycle defines the quantity called entropy. Entropy is related to the amount of disorder in a system and is a statistical description of the disorder. The second law of thermodynamics is a statement that when heat is transferred from one system to another to do work, the entropy and the amount of disorder increases.

4
Impurities are Key

Every civilization since the beginning of recorded time has given great value to jewels as symbols of position and wealth. What makes gems one of nature's treasures is their dazzling colors. Twentieth-century science has shown that the colors result from trace impurities in otherwise perfect crystals. In contrast to gems, the jewel of twentieth-century physics is the computer chip, where the basic building block of the electronics industry, the transistor, is made by adding trace amounts of dopants to ultra-pure silicon crystals. The last chapter discussed the removal of the random impurities that occur in materials when they are grown. This chapter discusses how the addition of trace amounts of specific elements, dopants, to quartz and to silicon gives them value as man-made gemstones and as the ubiquitous computer chips on which we depend.

Over millennia mankind discovered a myriad of different minerals, and the rarer, harder, and more beautiful minerals are called gems. When different gems were first found, they were classified by their external appearance—the symmetry and the color of the crystals—and they were given names like amethyst, carnelian, and jasper. In more modern times scientists determined that many gems had the same composition and structure as silica. They only differed from sand in their form and color and in turn in their impurities.

As far back as the second millennium BC the Babylonians made jewelry from agate, carnelian, onyx, and smoky quartz (Aruz *et al.* 2008). The Bible lists the jewels attached to the embroidered ceremonial cloth, the breastplate, worn by the Jewish high priest. There are twelve jewels representing the twelve tribes of Israel (Exodus 28, 15–21 and 38, 8–14), and five of them: amethyst, agate, jasper, onyx, and sardius are all derived from sand. The book of Revelations (21, 19–21) lists the jewels that will adorn the foundations of the walls of the new Jerusalem. These include amethyst, carnelian, chrysoprase, and jasper. In modern times birthstones are designated for each month. Among the sand-related birthstones are amethyst (February), bloodstone (March), sardonyx (August), and opal (October).

66 | SAND AND SILICON

To illustrate how twentieth-century physics reveals the similarities and the differences between these sand based gems, consider the collection of quartz gems in Fig. 4.1. Although they are different, x-ray diffraction studies show that each gem has the same crystal structure as sand (α-quartz). They differ in their color. Rock crystal is colorless whereas amethyst is violet-brown and citrine is yellow. These differences come from the impurities that are incorporated into quartz.

If an amethyst appears to our eye to be purple, then the other colors in visible light are absorbed by the crystal. As will be discussed in Chapter 5, light is quantized into units of energy called photons, and the energy of a photon is inversely proportional to its wavelength. A photon is absorbed in a material by transferring all of its energy to an electron that in turn is promoted to a higher energy or excited state. In the quantum theory of solids that was developed in the 1930s, the energy of an electron in a crystal can not have any arbitrary value. There are allowed values of energy that fall into what are called "allowed energy bands," and these are separated by disallowed energies that are called "energy gaps."

In an insulator like quartz, the electrons fill all of the allowed energy states in the lowest energy bands. In order to be absorbed, a photon must have sufficient energy to promote an electron from within an occupied band to one of the unoccupied bands. The energy gap in quartz is 10.2eV (the electronvolt, eV, is the kinetic energy gained by a free electron that is accelerated through an electric potential difference of one

Figure 4.1 The colors of gemstones come from small amounts of impurities. Rock crystal (lower right) is quartz containing no impurities and is clear. The other quartz gemstones include smoky quartz, amethyst, rose quartz, citrine, and smoky quartz. Amethyst contains iron impurities and citrine contains iron oxide. (Reproduced from Post (1997) with the permission of the Smithsonian Museum.)

volt), and this corresponds to a wavelength of 122 nm (a nanometer (nm) is one billionth of a meter). Visible light is composed of light with wavelengths that are between 380 nm and 760 nm. As these wavelengths are longer than 122 nm, pure quartz doesn't absorb visible light.

Rock crystal is colorless because it is pure quartz. The colors of amethyst and citrine come from small amounts of iron impurities that were incorporated into the crystals during growth. These impurities can absorb visible light because their energy bands are closer together, and their energy gaps correspond to wavelengths that overlap the wavelengths of visible light.

The iron is incorporated into quartz in three ways. The iron atoms can replace a silicon atom in a crystal of quartz (substitutional sites) or they can reside in the tunnels of the quartz structure (interstitial sites). Finally, there can be microscopic regions of iron oxide distributed inside the quartz crystal. Each of these different types of iron sites absorbs light in different wavelength regions (Rossman 1994).

In amethyst the iron impurities occupy both substitutional and interstitial sites, but the amethyst crystal has to be exposed to ionizing radiation (x-rays or gamma rays) in order to absorb light. In naturally occurring amethyst, the radiation comes from cosmic rays and the radioactive decay of neighboring radioactive minerals during growth. Synthetic amethyst crystals are directly irradiated during production.

The irradiation leads to a complicated rearrangement of the electrons between the iron impurities that occupy lattice sites or are in between lattice sites. The resulting so-called color centers absorb light at ranges of wavelengths that are centered in bands around 357 nm, 545 nm, and 950 nm. As a result, the wavelengths in between these bands (wavelengths around 460 nm and 720 nm) are not absorbed, giving amethyst a violet-brown color.

With time amethysts lose their color because the color centers become deactivated. Heating amethysts to several hundred degrees centigrade also results in a loss of the violet-brown color and often leads to the yellow color found in citrine.

In citrine, the iron impurities are microscopic particles of iron oxide that absorb light at wavelengths below approximately 500 nm (the energy gap in iron oxide) giving citrine a yellow color.

It is these subtle differences in the electronic structure of the iron impurities in quartz that give amethyst and citrine their color. This is a simplified description, but it illustrates how the quantum mechanics of the absorption and emission of light from the impurity sites gives these quartz-based gems their color.

It is the impurities that are the key when forming gemstones. Nature does an exquisite job distributing the impurities so as to give them an array of colors. In the twentieth century, scientists have learned how to grow a wide variety of synthetic gems by controlling the concentration of impurities.

Let's shift from the absorption of light in quartz-based gems to the electrical properties of silicon, and discuss why the electrical conductivity is extremely sensitive to small amounts of dopants. The concentration of impurities in ultra-pure silicon is measured in parts per billion. The concentration of dopants that are added to silicon to make the transistors in computer chips is in the range of parts per hundred thousand to parts per million. These impurities determine the electrical conductivity.

The electrical conductivity of a semiconductor is between an insulator like quartz and a metal like copper. To conduct electricity, there must be energy bands that are partially filled so that electrons are free to move through the crystal. It is convenient to consider just two bands, called the valence band and the conduction band, that are separated in energy by an energy gap. If the valence band is full and the conduction band is empty, then the material is an insulator and doesn't conduct electricity. If there are partially filled bands, then the material is a metal and conducts electricity. As discussed in Box 4.1, in a semiconductor the energy gap is small enough that a few electrons are promoted from the valence band to the conduction band and these electrons can move through the crystal. The promoted electrons leave behind empty states called holes in the valence band into which other electrons in the valence band can hop, also enabling a current to flow through the crystal. In a gem or in a solar cell, electrons are promoted by absorbing light of the appropriate energy. In the absence of light, electrons are promoted by absorbing some of the energy that resides in the vibrations of the atoms in a crystal.

The electrical conductivity of a semiconductor is extremely sensitive to certain added elements. A silicon atom has four valence electrons. If a phosphorous atom with five valence electrons is added to silicon, then the extra electron goes into the conduction band. Similarly, if a boron atom with three valence electrons is added to silicon, then there will be a hole in the valence band. Adding phosphorous or boron to silicon is referred to as doping. Adding phosphorous leads to what is called *n*-type silicon, and adding boron leads to *p*-type silicon. In each case, the electrical conductivity of the doped silicon will increase in proportion to the number of phosphorous or boron atoms that are diffused into a silicon crystal. The modern electronics industry is based on devices made by doping silicon to change its conductivity.

Furthermore, semiconductor electronics is built around what is called the *p–n* junction that results when a single crystal of silicon has an interface that separates regions that are *p*-doped from regions that are *n*-doped. The *p–n* junction is a fundamental component of transistors, solar cells, light emitting diodes (LEDs), semiconducting lasers, and photodiodes. The physics behind the *p–n* junction is described in Box 4.1.

The chance discovery of the *p–n* junction in 1940 grew out of attempts to improve rectifiers. These are devices to convert an alternating current into a direct current, and they had been made from copper with an oxide coat since the 1920s. A metal contact is pushed into an overlaying oxide and current flows from the point contact through the copper oxide to the copper. Silicon point contact rectifiers operate at the higher microwave frequencies needed for radar. A tungsten wire made contact with a silicon plate, but it was hard to make reliable contacts because the impurity distribution in the silicon that was available at the time could not be controlled.

In 1940, Russell Ohl at Bell Labs was trying to make more reliable silicon rectifiers. He had prepared an ingot of silicon by melting it in a silica crucible and then slowly cooling it. The illustration of his experiment, taken from his patent, is shown on the left of Fig. 4.2. The silicon solidified starting from the bottom of the crucible. With hindsight it is clear that the lighter impurities such as boron had segregated to the top of the ingot while the heavier phosphorous impurities had segregated to the bottom. This produced an interface where the doping changed from boron to phosphorous. As boron and phosphorous have one less and one more electron than silicon respectively, the interface formed a *p–n* junction in the middle of the ingot.

Ohl cut a piece of silicon out of the ingot as illustrated on the right of Fig. 4.2. Much to his surprise, he found that when he made point contacts in different places along the silicon, the rectifier changed polarity. The polarity was positive when the contact was made to what was later found to be the *p*-doped region, and it was negative when the contact was made to the *n*-doped region.

Ohl also found that when the *p–n* junction was illuminated with light, a current flowed through the crystal. This latter discovery was the precursor to the silicon solar cell, which will be discussed in Chapter 5 (Ohl 1948).

At the time the concentrations of the impurities in silicon were too small to measure, so Ohl and his colleagues were guided by experience and intuition. Henry Theuerer discovered that when boron was added to

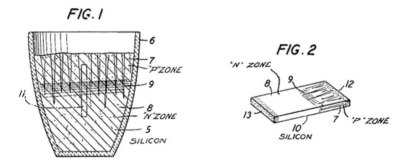

Figure 4.2 The discovery of the *p–n* junction. As the silicon in the crucible solidified, the impurities separated to give a *p*-doped zone and an *n*-doped zone on the top and bottom respectively. The slice cut from the middle of the ingot of silicon had a *p–n* junction, as illustrated on the right where the polarity of the point contact rectifier changed from positive to negative. (From US Patent 2,402,662.)

the melt, the *p–n* junction moved further down the ingot, indicating that the impurity at the top was boron. He guessed that the impurity in the bottom part of the ingot was phosphorous because he smelled a trace of phosphine when the crucible containing the ingot was opened. Russell Ohl and Jack Scaff named the top region "*p*-type" for positive and the bottom region "*n*-type" for negative based on the polarity of the point contact rectifiers in the different regions, and the *p–n* junction was born (Millman 1983).

> **BOX 4.1 Semiconductors and the *p–n* Junction**
>
> To illustrate the electrical behavior of a semiconductor, we construct a hypothetical model of a two-dimensional material with one electron per atom. This is analogous to the board game Chinese checkers as illustrated in Fig. 4.3. In Chinese checkers the board has a regular array of hemispherical depressions, and the marbles are moved from one depression to another. In the hypothetical model of a semiconductor, one board represents the valence band and another board represents the conduction band. The depressions on each board are the positions of the nuclei of the atoms, and the marbles are the electrons. The separation between the boards in the vertical direction is the amount of energy needed to promote an electron from the valence band to the conduction band.
>
> At a temperature of absolute zero all of the marbles (electrons) fill the valence band (bottom checker board), as shown in the top left of Fig. 4.3, because there is no thermal energy available to promote electrons to the conduction band. At room temperature there is enough thermal energy to promote a few electrons to the conduction band (top right checker board) leaving behind a corresponding number of empty holes in the valence band. The figure is a snapshot in time of the positions of the electrons and the holes. The electrons in both the valence and conduction band can hop from one empty depression to another.
>
> This is an idealized model because in quantum mechanics the electrons occupy atomic orbitals and the positions of the electrons are defined in terms of probabilities instead of absolute positions. The checkerboard model gives a simplified picture from which to visualize the properties of a semiconductor.
>
> As with Chinese checkers the object is to get the marbles from one side of the board to the other. The electrical conductivity is equal to the current (number of electrons moving across the board per unit time) divided by the voltage (driving force on the electrons). To mimic the motion of electrons, assume that an electron has a probability of hopping to an adjacent empty site and that if the checker board is tilted from the horizontal, there is a bias to hop toward the lower side of the board.
>
> At a temperature of absolute zero (Fig. 4.3), there are no empty places in the valence band to allow the electrons to move so the conductivity is zero. At room temperature, if the boards are tilted to the right, the marbles can hop to the right in both bands and the conductivity has some finite value. In the valence band the electrons hop preferentially to the right, and the empty depressions appear to move to the left. It is customary to say that the empty depression or hole is the same as a positive charge. It is appropriately called "hole" conduction.

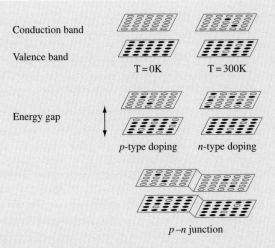

Figure 4.3 Hypothetical model of a semiconductor with one valence electron per atom. (top) Electrons are promoted from the valence band to the conduction band as the temperature increases. (middle) The presence of extra holes in p-doped material and extra electrons in n-doped material. (bottom) A p–n junction is formed when the two are brought together. Electrons on the right fill holes on the left leaving no free electrons at the junction. This flow of charge results in a barrier to further flow of electrons and holes (the barrier is represented by the vertical displacement of the valence and conduction bands at the interface). The p–n junction acts as a rectifier in which current can flow if the p-side is made positive, thereby reducing the barrier, and in which no current flows if the n-side is made positive, thereby increasing the barrier. The p–n junction is the key component in transistors, solar cells, LEDs, and semiconductor lasers.

The electrical conductivity of a semiconductor changes rapidly with the addition of impurities that add electrons or holes. If the number of impurities increases, then the conductivity increases. In the hypothetical model, there are two types of impurity atoms: those that have no electron and those that have two electrons. If the impurity has no electron, there will be a hole in the valence band, and the conductivity will increase because the hole can move through the crystal. This is illustrated in our checker board model (middle left) where, in addition to the holes produced by thermal excitation, there are additional holes produced by the doping. An impurity atom having two electrons contributes one electron to the valence band and one electron to the conduction band that can move through the crystal (middle right of Fig. 4.3).

Next consider a crystal (checker board) that is p-doped on the left and n-doped on the right (bottom of Fig. 4.3). Electrons will diffuse from the region with higher electron concentration (n-doped) to that with the lower electron concentration (p-doped) where they recombine with holes as illustrated in the bottom of Fig. 4.2. Recombination is the process where an electron in the conduction band loses energy and fills a hole in the valence band.

(*cont.*)

> **BOX 4.1** *Continued*
>
> This preferential diffusion and recombination can't go on too long because there will be a net build up of charge near to the interface between the *n*- and *p*-doped regions. The *n*-doped region has a net positive charge because it has lost negatively charged electrons, and the *p*-doped region has a negative charge because it has gained electrons. This imbalance creates an opposing electric field that is a barrier to further charge transport. The barrier is represented by a vertical displacement of the checker boards at the junction. In the so-called depletion layer at the junction between the *n*- and *p*-doped regions, there are no electrons in the conduction band or holes in the valence band. Outside of the depletion layer there will be the normal distribution of electrons and holes in the *n*- and *p*-doped regions.
>
> The *p–n* junction has a very important property; it is a rectifier. If a voltage is applied across the sample such that the *p*-side is positive and the *n*-side negative (forward biased), then the barrier at the junction is decreased. It is easier for electrons to overcome the barrier. Electrons can flow across the sample and the conductivity increases rapidly with increasing applied voltage. Conversely, if the voltage is reversed (reversed biased), then the barrier increases and the conductivity is low. (The vertical axis in Fig. 4.3 is the relative energy of the electrons in the valence and conduction bands. A positive bias voltage results in a decrease in the energy levels in the *p*-doped material relative to the *n*-doped material.) The *p–n* junction is called a diode, meaning that current can flow in one direction but not in the other.

It was against this background that the transistor was invented in 1947 at Bell Labs. The transistor was the solid state replacement for the vacuum tube that had been a key component of radio and telecommunications for half a century. William Shockley, John Bardeen, and Walter Brattain received the Nobel Prize in 1956 for the invention of the transistor. The evolution from the original point contact transistor (Bardeen and Brattain 1948), to the junction transistor (Shockley 1948), to the complementary metal oxide semiconductor (CMOS) transistor (Wanlass 1963), has been described many times (Orton 2004; Riordan and Hoddeson 1997; Reid 1958; and http://www.computerhistory.org/semiconductor/timeline). A brief review of this evolution is given to illustrate how the controlled addition of minute quantities of dopants in silicon is the foundation of the electronics industry.

For almost three-quarters of the twentieth century, Bell Telephone Laboratories, Inc. was one of the leading industrial research laboratories in the world. The Bell System was composed of AT&T, the regional telephone companies, and the Western Electric Company that was its manufacturing arm. The management at AT&T realized that it needed a separate subsidiary to do research to improve the telephone system and to provide patent protection for its telephone monopoly. In 1925 the research section of the Western Electric Company was separated from the parent company to form Bell Telephone Laboratories, Inc. that was a wholly owned subsidiary of AT&T and Western Electric. In 1996 Bell Labs and the remaining parts of Western Electric were recombined and spun off as a separate company named Lucent Technologies in the aftermath of the breakup of the Bell System. During the intervening 71 years Bell Labs was a scientific and engineering powerhouse. It produced over 30,000 patents and 7 Nobel Prizes. The first prize was for establishing the wave-particle duality of the electron (see Box 6.1) and the most recent was for the CCD detector (see Chapter 5). Perhaps the most well known of the Bell Labs Nobel Prizes is for the invention of the transistor.

After the Second World War it was clear that a solid state replacement had to be found for the vacuum tube. It was not robust enough, and it relied on heated filaments that were prone to burning up. It was also clear that the telecommunications industry was going to expand rapidly after the war. This meant that vastly increased numbers of smaller, lower power, solid state amplifiers and switches were going to be needed to operate the telephone network.

Bell Labs set up a research department to study the physics of semiconductors, headed by William Shockley, with the hope that a new concept for a solid state amplifier would be found. Among the members of his group were Walter Brattain and John Bardeen. This was a formidable team.

The branch of physics called solid state physics (now called condensed matter physics) started with the application of quantum mechanics to understand and calculate the properties of metals, semiconductors, and insulators in the early 1930s. As the field evolved there was a natural division of labor between what are called theorists and experimentalists. The theorist has to calculate the properties of an ensemble of 10^{23} atoms and take into account all the interactions between the electrons and the nuclei. It is possible to understand semiconductors in the broad sense that is illustrated in Fig. 4.3, but to actually predict any particular property quantitatively requires complex calculations. Even with present day computing power, this is difficult to do. Theorists have to make the right approximations to bring out the key underlying interactions that lead to a particular property.

By contrast, experimentalists deal with the complexities of real materials. They have to purify and characterize new materials (see Chapter 3), control the properties of the surfaces of crystals, modify the properties of materials by adding dopants, and make a myriad of sophisticated measurements to reveal the atomic and electronic properties of materials.

Walter Brattain was the experimentalist in the transistor triumvirate. He had grown up on a ranch in Washington and graduated from Whitman College in Oregon before receiving his Ph.D. from the University of Minnesota, where he studied the interaction of beams of electrons with the surface of mercury. After a year at the National Bureau of Standards, he became one of the early employees of Bell Labs—joining the company in 1929. He made notable contributions to the understanding of the surfaces of materials and was involved in improving the copper oxide and silicon rectifiers that were discussed above. He had exactly the talents that were needed as the team tried one experiment after another to make a solid state amplifier.

William Shockley and John Bardeen were the theorists. Shockley had grown up in California and gone to Cal Tech. He joined Bell Labs in 1936 after graduating from MIT where he did one of the first calculations of the electronic band structure of a simple material—table salt—sodium chloride. He began a quest for a solid state amplifier, but the shadow of the impending Second World War led to a hiatus in this work.

Shockley's biographer, Joel Shurkin, relates a story that demonstrates the breadth of Shockley's intellect. The president of Bell Labs, Mervin Kelly, asked Shockley and Jim Fisk, who was later to become President of Bell Labs, "Can nuclear energy be made available by the fission process?" In two months, they had designed a nuclear reactor and applied for a patent. Everything dealing with nuclear power or potential nuclear weapons was immediately classified, but at the end of the war when the records were opened to process patent applications it turned out that the Shockley–Fisk patent had the earliest date (Shurkin 2006).

John Bardeen was the second theorist on the team. He, like J. Willard Gibbs, was a quiet genius. Someone who didn't say much but when he did it was profound. He is the only person to win two Nobel Prizes in Physics—one for the transistor and one for the theory of superconductivity. He received a degree in electrical engineering from the University of Wisconsin and worked for three years for Gulf Oil Company before beginning graduate work at Princeton. Princeton was one of the centers of the evolving field of solid state physics, with many of its founding fathers, like Eugene Wigner and Fred Seitz, on the faculty. For his thesis Bardeen calculated the work function of sodium metal. The work function is the amount of energy that is needed to remove an electron from the surface of a crystal. He was a Junior Fellow at Harvard in 1935 before becoming a professor at the University of Minnesota. During the war he worked at the Naval Ordinance Laboratory on magnetic mines and torpedoes before joining Bell Labs in 1945 (Hoddeson and Daitch 2002).

It is interesting to note how many of the famous scientists that we have discussed worked on the detection of and the destruction of submarines. In the First World War, Paul Langevin (Chapter 1), William Bragg (Chapter 2), and Ernest Rutherford (Box 4.1) all tried to develop methods to detect submarines. In the Second World War, William Shockley used operations research to find the optimum strategies to find and destroy submarines while John Bardeen worked on mines and torpedoes.

The team assembled to search for a solid state amplifier was ideally suited to the task. Brattain had experience in surface science, Shockley had applied quantum theory to the electronic structure of crystals, and Bardeen had calculated the electronic properties of surfaces.

Shockley had an idea for a field effect amplifier in which a strong electric field near the surface of a crystal would lead to an increased flow of electrons in the crystal. It worked as an amplifier because the flow of electrons was controlled by the strength of the applied electric field. Shockley calculated that there should be a sizable amplification, but when Walter Brattain and Russell Ohl tried to construct the device it didn't work. Bardeen realized what the problem was. It was assumed that the electrons at the surface would be just as free to move as those inside the crystal, but in fact, the electrons were being trapped in what are now called surface states so the conductivity didn't increase with increasing electric field.

Bardeen and Brattain tried a different approach based on the principle behind the triode vacuum tube amplifier that was the backbone of the communications industry for a quarter of a century. In a triode the flow of electrons from the cathode to the anode is controlled by a grid of wires placed between them. Bardeen and Brattain had the idea of putting a second point contact (the emitter) very close to the primary point contact (the collector) in a point contact rectifier. This, they hoped, would act like a grid to modify the flow of current from the gold contact on the top of a germanium crystal to the base electrode on the bottom of the crystal.

Over a two month period from November to December 1947, they struggled both to construct the device and to understand the associated flow of electrons and holes between the three different contacts. Brattain came up with a design that led to amplification and overcame the surface states, and Bardeen showed that the operation of the device relied on the flow of holes from the emitter to the collector. The holes controlled the flow of the electrons from the collector to the base. This is the fundamental new idea behind the point contact transistor. John Bardeen and Walter Brattain had demonstrated a solid state amplifier and the modern electronics industry was born.

No sooner had Bardeen and Brattain achieved their goal than the dream team started to disintegrate over the question of credit for the invention. Bell Labs, like most large organizations, had a hierarchal structure. Shockley was the Department Head who had initiated the search for a solid state amplifier. However, the invention of the point contact transistor was entirely the work of Bardeen and Brattain, and in fact, the patent was assigned to them by the Bell Labs lawyers. The official Bell Labs position was and is that the transistor was invented by all three. In keeping with the hierarchal structure, the official photograph shows Shockley, the Department Head, looking through the microscope at the invention while Bardeen and Brattain look on.

Relations within the group broke down. Shockley realized, correctly, that the point contact transistor, although a proof of principle, would be hard to manufacture. He wanted to develop his own transistor, partly to establish his place in the discovery of the transistor. Working with Morgan Sparks and Gordon Teal in his group, he set about developing a transistor that operated entirely within a crystal of germanium rather than relying on fragile point contacts on the surface of the crystal. They invented the junction transistor described below (Riordan and Hoddeson 1997 and Shurkin 2006).

Bardeen wrote to the President of Bell Labs that the working relations in the group had become untenable, and he left to become a professor of physics at the University of Illinois. With colleagues Bob Schreiffer and Leon Cooper, he developed the theory of superconductivity that had eluded other theorists since its discovery in 1913. For that work he shared in a second Nobel Prize in physics.

Shockley left Bell Labs to form Shockley Semiconductor in California, partly because he didn't feel that the company had rewarded him sufficiently. Shockley's move to California gave birth quite literally to Silicon Valley. His management style, however, led to a second breakup of an outstanding research team as eight of the team left Shockley's company to form Fairchild Semiconductors in 1957. The latter company was very successful, and subsequently Robert Noyce and Gordon Moore, who were part of the eight, went on to form Intel in 1968. After the demise of Shockley Semiconductor, Shockley became interested in eugenics and possible differences in intelligence among different races, leading to his isolation from the scientific community in his later years.

What had started as an example of enlightened technical management at Bell Labs in assembling an outstanding team and then letting them loose to solve a problem critical to future telecommunications, ended with human competition and egos leading to the dissolution of the team. Nevertheless, the invention of the transistor sparked the evolution of the modern electronics industry on which we rely today. Let's follow the key developments of the industry beginning with the junction transistor.

The junction transistor, which was invented by Shockley, is made by putting two p–n junctions next to each other to form an n–p–n (or p–n–p) structure. An early n–p–n junction transistor made at Bell Labs out of germanium is shown in Fig. 4.4. Dopants are diffused into the single crystal to produce a thin p-doped region in between the n-doped regions at each end of the bar. Thin wires make contact with each region. When a voltage is applied to the intermediate p-doped region (the base), both junctions are biased. The n–p junction is forward biased, and the p–n junction is reversed biased. As a result, the emitter injects electrons into the p-type base region, and the electrons are driven into the n-type collector region resulting in a current. The junction transistor functions as an amplifier with small changes in the base voltage resulting in large or amplified currents as electrons move more easily from the emitter to the collector with increasing base voltage.

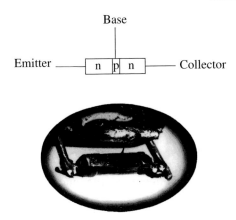

Figure 4.4 Early germanium junction transistor. A thin section in the middle of the n-doped bar is p-doped so as to make a n–p–n junction transistor. A voltage applied to the base increases the flow of electrons from the emitter to the collector leading to amplification. Reproduced with permission of Alcatel-Lucent USA Inc.

The second major type of transistor is the field effect transistor (FET). It is compared with a junction transistor in Fig. 4.5. (Both transistors in the figure are made from silicon using planar technology that is described below.) In an FET a gate electrode draws electrons or holes to the surface and provides a channel for conduction from one p–n junction to the other. Superficially the FET appears to be similar to a junction transistor, but there is a fundamental difference. Electrical contact is not made to the middle region in an FET. Instead an electric field is applied by what is called the gate electrode. The gate is separated from the Si by an insulating oxide layer, thus the name metal-oxide-semiconductor field effect transistor (MOSFET). n–p–n and p–n–p field effect transistors are referred to as NMOS and PMOS transistors respectively.

In 1926, Julius Lilienfeld had patented the idea of controlling the current flowing through a film by applying an electric field, but he never successfully demonstrated a field effect transistor (Lilienfeld 1926). Early attempts, especially by Shockley, to make an FET were frustrated by the fact that the electrons drawn to the surface under the electrode were trapped at the surface of the silicon at various types of imperfections. The development of the oxide passivating layer (discussed below) minimized the trapping of electrons while also acting as an insulating layer on which to deposit the gate electrode. The first practical FET was developed by Dawon Kahng at Bell Labs (Kahng 1960).

Combinations of NMOS and PMOS transistors are used to make the logic circuits that are the heart and soul of computers. In 1963, Frank Winlass of Fairchild designed a circuit using complementary pairs of NMOS and PMOS transistors that use almost zero stand-by power. The ability to a make low power-consuming circuit is key to commercial

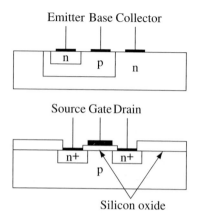

Figure 4.5 Transistors made using planar technology have the different doped regions near the surface of a doped silicon crystal. In the junction transistor (top) a voltage on the base controls the current flow from the emitter to the collector. In the N metal-oxide-semiconductor field effect transistor (N-MOSFET) (bottom), the gate is separated from the silicon by a layer of silicon oxide, and a voltage on the gate pulls electrons to the surface under the oxide to increase the current from the source to the drain.

applications like portable computers and cell phones that rely on battery power. The resulting circuit is called complementary metal oxide semiconductor or CMOS, and it dominates silicon technology.

The large scale manufacture of transistors stems from the invention of planar technology by J. R. Hoerni at Fairchild Semiconductor (Hoerni 1959). In planar technology all the components are made on (or near) the surface of a semiconductor single crystal. In a junction transistor (Fig. 4.4) a small region of an *n*-doped single crystal is *p*-doped. Enough of the *p*-dopant is diffused into the *n*-type material to compensate or neutralize the *n*-doping and make the region *p*-type. Finally a smaller region within the *p*-doped area is doped back to *n*-type. In this way there is an *n*-doped region nested in a *p*-doped region that in turn is nested in the bulk *n*-doped region.

Today's silicon technology is based on the realization by Jack Kilby at Texas Instruments that all the components needed for an electronic circuit—resistors, capacitors, and transistors—can be manufactured on the same piece of semiconductor (Kilby 1959). Appropriate layers are deposited onto the surface of the single crystal, and impurities are diffused into the appropriate areas to manufacture semiconductor devices.

Almost simultaneously in 1959, Robert Noyce at Fairchild Semiconductor produced the first integrated circuit using planar technology. The different circuit elements were grown on a silicon crystal and then connected by wires. As the technology evolved, the wires were replaced with metal contacts that were deposited on the protective oxide film discussed below.

The integrated circuits invented by Kilby and Noyce addressed the so-called tyranny of numbers. How can circuits needing millions of components be manufactured at a reasonable cost? The integrated circuit made it possible to produce all the components in parallel on a single crystal of silicon and to make many copies of the circuit on the same crystal at the same time. Jack Kilby shared the Nobel Prize in 2000 (unfortunately, Robert Noyce had passed away before the Nobel Committee made the award).

In only 12 years the field of semiconductor physics had gone from the first demonstration of the transistor, which proved that it was possible to develop a solid state amplifier to replace the vacuum tube, to the integrated circuit that revolutionized the electronics industry. As John Orton points out in his account of *The Story of Semiconductors*, timing is everything (Orton 2004, p. 99). The production of integrated circuits was not originally cost competitive with the traditional component by component manufacture of electronics, and the nascent semiconductor industry was struggling in the 1950s. However, Sputnik was launched in 1957, and two years later President Kennedy made it the mission of the United States to put a man on the moon by the end of the following decade. Although the initial research at Bell Labs had been funded by AT&T because of the need to replace the vacuum tube in the telephone system, it was government funding that supported the development of silicon technology in its early years because of the need for lightweight compact electronics for space and defense programs.

Sand in the form of silicon is the cornerstone of today's electronics industry, but equally important is the role played by sand in the form of silicon dioxide. In order for silicon technology to work, the surfaces of the doped silicon regions are pacified by covering them with oxide so the electrons don't get trapped in imperfections at the surface. At the same time these tiny circuits are very fragile, and an oxide film protects them from damage and external contamination. Until the most recent CMOS circuits, silicon dioxide (sand) was used to separate the gate electrode from the silicon (Fig. 4.5). Of equal importance is the use of silicon oxide films in the manufacturing process to define the areas that have to be selectively doped.

The discovery of the remarkable properties of silicon oxide films was the result of an accident. In the early days of transistor development, the dopants were diffused into the appropriate regions. Hydrogen gas containing the dopant was passed over the silicon. The silicon in turn was held at a high temperature to enhance the diffusion of the dopants into the silicon. One day in 1955 Lincoln Derick and Carl Frosch at Bell Labs were running a diffusion furnace when the hydrogen gas accidentally caught fire. When hydrogen burns it forms water, and the heated silicon surface immediately reacted with the water to produce a film of silicon oxide. The oxide turned out to be a strong, stable, pinhole-free film. This is indeed fortuitous because few materials form pinhole-free oxide layers.

Derick and Frosch went on to show how the oxide film could be used to define the areas on the silicon that were to be doped. They determined that many of the common dopants did not diffuse through the oxide and enter the silicon. The areas that are to be doped can be defined by etching off the oxide as described below (Derick and Frosch 1956).

This brief review illustrates how the careful placement of dopants in the right places and in the right concentrations in ultra-pure silicon is used to make today's integrated circuits. In addition to the development of the physics of semiconductors, it is the control of silicon and silicon oxide that has made silicon technology possible. As the electronics industry continues to follow Moore's Law, with the number of transistors in an integrated circuit doubling every 18 months, the size of each individual transistor gets smaller each year. The INTEL 4004 Microprocessor that was introduced in 1971 had 2,300 transistors (Fig. 4.6). The INTEL® Pentium® 4 Processor introduced in 2000 had 42,000,000 transistors. During that time the width of the gate electrode shrank from 10 micrometers to 0.18 micrometers. One of the engineering marvels of the twentieth century is the progression from the early primitive junction transistor in Fig. 4.3 to the microprocessors in Fig. 4.6. This progression is reflected in the silicon and silicon oxide that figuratively start as little grains of sand.

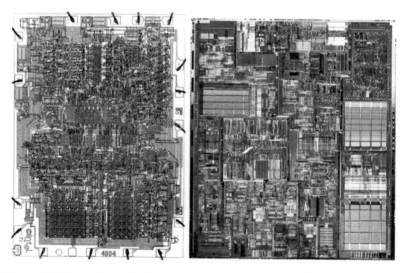

Figure 4.6 Comparison of microprocessors made in 1971 and 2000 showing the dramatic increase in the number of transistors, the clock speed, and the decrease in the gate width. (left) INTEL 4004 Microprocessor, 2,300 transistors, 108 KHz, 10 μ and (right) INTEL® Pentium® 4 Processor, 42 M transistors, 1.5 GHz, 0.18 μ. Reproduced with the permission of Intel Corporation.

How is it possible to create integrated circuits containing millions of transistors, capacitors, and resistors on penny-sized slices of silicon? In a greatly simplified view of the manufacturing process there are two principle elements. First is the need to define the areas on the surface of the silicon crystal that are to be doped or covered with a protective layer of oxide or a metal gate or contact. Second is the need to add dopants in the right concentration and at the right depth into the silicon. The first is accomplished using photolithography. The second is accomplished using ion implantation.

Lithography has had a venerable place in the world of printing for over 200 years. Many of us have lithographs hanging on our walls. The original process involves drawing a picture on a stone tablet using wax and then wetting the stone with water. The water is repelled by the wax. When the ink is applied, it only goes to the places that are covered by the wax, thereby allowing the image to be transferred from the stone to paper when one is pressed against the other.

With the development of photography, the stone plate is replaced by a metal plate coated with an emulsion in place of the wax. The plate is exposed to light shown through a film containing the picture to be reproduced. When the emulsion is developed, the emulsion under the image has not been exposed to light and it remains on the plate. When the ink is applied, it goes to the places covered by the emulsion as in the original process.

Photolithography was adapted to manufacture electronic circuits by Jules Andrus at Bell Labs (Andrus 1957). The silicon wafer is coated with a light sensitive film called a photoresist. A mask is positioned above the wafer that is transparent to light in the regions to be implanted or where contacts are to be made. The mask is then exposed to light and subsequently the photoresist is developed exposing the appropriate areas.

The process has to use light with a shorter wavelength than visible light because the resolution of photolithography is limited by the properties of light. The resolution of optical instruments is limited to distances that correspond to roughly the wavelength of the light that is used to observe the object. To obtain sharp patterns of objects with dimensions of tenths of a micrometer, like the gate electrodes in field effect transistors, it is necessary to use ultraviolet light. Currently the industry uses ultraviolet light with a wavelength of 0.193 micrometers.

The resolution also depends on the distance between the mask and the film. To reproduce feature sizes of tenths of a micrometer, the mask has to be brought very close to the surface of the film to minimize blurring. The present technology using 0.193 micrometer UV-light has been very successful, but as the feature size moves well below a tenth of a micrometer, it will be necessary to use shorter wavelengths in the soft x-ray region. This will require a change from proximity printing, where the mask is placed very close to the silicon wafer, to optics which project a reduced image of the mask onto the wafer.

Let us follow the first few steps that have to be taken to manufacture the N-MOSFET shown in Fig. 4.5. Starting with the p-doped substrate, the first step is to form the n-doped regions. To do this, a layer of silicon oxide is grown on the surface of the silicon crystal. Next, a layer of a UV sensitive photographic emulsion or resist is applied to the wafer. A mask that is transparent to UV light in the regions where the n-doping is needed is optically aligned on the wafer. The wafer is exposed to UV light and then developed. The regions to be implanted are now exposed, except for the oxide layer that is removed using a chemical etch. This exposes the regions of the p-doped silicon that are to be n-doped. The rest of the resist is removed leaving the wafer with a protective layer of silicon oxide everywhere except in the regions to be implanted.

As described below, ion implantation is used to dope the exposed region, while the silicon oxide stops or masks the dopants from reaching the rest of the silicon. Enough phosphorous is implanted to compensate for the boron dopant that was used to make the p-type substrate, and then more phosphorous is implanted to make the regions n-type.

After completing the necessary doping, a similar sequence is used to define first the thin oxide layer of the metal-oxide-semiconductor device and the gate electrode. Finally, the process is repeated to deposit the electrical contacts used to form the N-MOSFET. In a computer chip containing both N-MOSFETs and P-MOSFETs, the whole sequence of steps may be repeated more than a dozen times depending on the circuit being manufactured.

The process of ion implantation is used to selectively dope regions of a silicon wafer. The first ion implantation experiments were done at Bell Labs in the 1950s. R. S. Ohl suggested that directing a beam of ions from an accelerator on the surface of a semiconductor crystal could modify the surface so as to enhance the diffusion of dopants into the crystal (Ohl 1950). Subsequently, Shockley described how ion implantation could be used directly to dope semiconductors (Shockley 1954). This is an attractive process because the exact quantity, position, and depth of dopant can be controlled. As discussed in Box 4.2, ion implantation builds on basic research aimed at understanding how charged particles interact with atoms in a material.

In order to implant silicon to make it into a p-type semiconductor, a beam of boron ions is directed at the wafer (Rubin and Poate 2003). Similarly, for an n-type semiconducting region a beam of phosphorous ions is used. An accelerator is used to produce a beam of positively charged boron or phosphorous ions. The beam passes through a magnet that bends it at 90 degrees. The strength of the magnetic field selects the ions with the right energy to implant at the desired depth as depicted in Fig. 4.7. The concentration of dopants is controlled by the number of ions per unit area that hit the silicon surface.

It is remarkable how experiments done a hundred years ago, to understand how alpha particles interact with matter and to prove that atoms have small positively charged nuclei at their centers, are the basis of the technology of ion implantation that is used in the electronics industry to dope semiconductors. Coupled with photolithography, ion implantation makes it possible to selectively dope millions of sub-micron-sized regions in a computer chip. Electrical engineers are able to design incredibly intricate circuits for computers, cell phones, and so on, but in the end it is the ability to selectively control the placement of dopants that is the critical factor in the production of our everyday electronics and computers.

BOX 4.2 The Structure of the Atom

Our understanding of the nucleus of the atom is largely based on experiments where various types of charged particles are accelerated to high velocities to hit a target or to collide with charged particles traveling in the opposite direction. Today, new discoveries in high energy and nuclear physics come from enormous accelerators like the Large Hadron Collider at CERN, the Tevatron at Fermi Lab, and the Relativistic Heavy Ion Collider at Brookhaven. These "big science" experiments use the latest in a progression of bigger and better accelerators that have been constructed over the course of the last century, but they stem from small yet elegant experiments done at the beginning of the century.

At the beginning of the 20th century, the atom was believed to be some kind of immutable ball. The existence of the nucleus and its structure was unknown. The discovery of x-rays, the electron, and radioactivity near the beginning of the twentieth century shattered this simplistic view. What were originally called alpha, beta, and gamma rays emanated from certain radioactive materials. Where did these different particles and rays come from?

The early experiments to elucidate the structure of the atom did not involve accelerators but beams of alpha particles. The Curies had discovered the element radium (Box 1.1). As radium decays to radon, it emits highly energetic alpha particles. A beam of alpha particles is produced by appropriate collimation. The first scientist to use beams of alpha particles to study the nucleus was Ernest Rutherford.

Rutherford was born on 30 August 1871 on a farm in New Zealand. He received his early training at Canterbury College in Christchurch and then won a prestigious fellowship to study at Cambridge University in England in 1895. After working with J. J. Thompson, who discovered the electron, Rutherford became a professor of physics at McGill University in Canada in 1899 (Reeves 2008).

At McGill, he collaborated with the chemist, Frederick Soddy, in a series of ground breaking experiments on radioactivity. First, they measured how radiation changed with time. Thorium emissions decreased to half their value in 60 seconds, and then the remaining radiation decreased to half of its value in the next 60 seconds and so on. Sixty seconds is the half-life of thorium.

Then, they determined that there were a series of decay processes. At the time the concept of isotopes had not been developed, but they could tell that

(*cont.*)

BOX 4.2 *Continued*

there were a series of discrete emanations, each with its own half life. Chemical studies showed that the different emanations came from substances that had the properties of other chemical elements like actinium, radon, bismuth, and lead. They established the transmutation of radioactive elements through a series of disintegrations. Rutherford received the Nobel Prize in Chemistry in 1908 for these discoveries. He quipped, when he received the Prize, that the speed of the transformations that he had observed was less than the speed with which the Nobel Committee had transformed him from a physicist to a chemist.

As a byproduct of a series of transmutations, Rutherford realized that he could determine the age of any given sample of radioactive material. He had determined the half-life of each of the decay products, and if he measured the strength of the radiation from each at any given time, then he could calculate backwards in time to when the rock had been formed and only contained the parent element. He had invented radioactive dating.

In 1907 Rutherford moved to the University of Manchester in England. It was here that he proved that alpha particles are helium atoms that have lost their two electrons. During this time he began to use beams of alpha particles to probe the structure of matter. He asked Ernest Marsden and Hans Geiger to study the scattering of alpha particles from a gold foil. (Geiger is the inventor of the familiar Geiger counter that is used to detect radiation.) They set up an elegant but simple experiment shown schematically in Fig. 4.7. A beam of alpha particles was defined by a collimator and directed toward a thin gold foil. Phosphorescent detectors, which consisted of a coating of zinc sulfide on a screen, were placed around the foil. When an alpha particle hit the screen, it produced a tiny flash of light that could be seen through a microscope in a darkened room. Marsden spent hours in the room observing where the flashes occurred on the screen. He found that most of the flashes were observed directly behind the foil indicating that the alpha particles had passed through the foil with either no deflection or only a small deflection. The surprising result was that a small number of flashes were observed at large angles—even in front of the foil. Ernest Rutherford later commented that "It was almost as incredible as if you fired a 15-inch shell at a piece of tissue paper and it came back to hit you." It was over a year before Rutherford grasped the full implications of the experiment and proposed a new model for the structure of the atom.

At that time, the conjecture was that the positive charge that balanced the negative charge of the electrons was distributed uniformly through the atom—the so-called pudding model. If this model was correct, then the alpha particles would just pass right through the foil. If, however, the positive charge is concentrated in what we now call the nucleus, then the alpha particle can be deflected by large angles. This is analogous to the cue ball hitting a pool ball directly and reversing direction.

To explain these observations, Rutherford proposed that atoms were not homogenous balls but were composed of a central very dense nucleus with the much smaller electrons distributed around it. He developed the theory for what is now called Rutherford scattering. In the theory the alpha particles approach the atom, and the positively charged alpha particle and positively charged nucleus repel each other. If the alpha particle only passes by the nucleus at a distance,

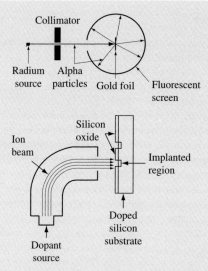

Figure 4.7 Comparison of the experiment of Marsden and Geiger, which was used to determine the existence of and the size of the nucleus of an atom, and ion implantation, which is used to dope semiconductors in the manufacture of computer chips. In the first experiment beams of alpha particles scatter off a thin gold foil and are detected as flashes of light on a fluorescent screen. In the second dopant ions are accelerated and then implanted in the appropriate regions of a silicon wafer to make transistors. The first experiment fits on a table top and the second fills a room.

then it is only slightly deflected. If the alpha particle comes close to the nucleus, then it can be deflected by a larger angle—even 180° to its original direction. So the tedious experiment of counting flashes of light for hours on end in a dark room led to the current model of the atom with a small positively charged nucleus at its center (Reeves 2008 and Lightman 2005).

The model was not yet complete because one had to explain what the electrons were doing. Why didn't the negatively charged electrons slowly lose energy and spiral into the positively charged nucleus? This problem was solved by Niels Bohr, who came to work in Rutherford's laboratory in 1911, and he won the Nobel Prize in 1920 for this work. He proposed that the electrons moved in fixed orbits without radiating energy and that the orbits are quantized with their angular momentum being multiples of Planck's constant divided by 2π. As discussed above and in Chapter 5, electrons can only move from one orbit to another by absorbing energy that equals the difference between the energy of adjacent orbits. Moreover, the number of electrons in each orbit is limited by the symmetry of the orbits, and electrons can only be promoted into orbits that are not fully occupied. Bohr's model for the electronic structure of the atom is the foundation on which the modern quantum mechanical description of atomic structure was built in the 1920s by Erwin Schrödinger and Werner Heisenberg.

(cont.)

BOX 4.2 *Continued*

The studies outlined above formed the basis for understanding how beams of positively charged alpha particles interact with the nucleus. The alpha particles also interact with the electrons in the atom. The alpha particles have much more kinetic energy and are much heavier than the electrons. As a result there is only a small change in the alpha particle's momentum. One can think of the sea of electrons as being similar to a viscous fluid that provides a general drag on the alpha particles, causing them to slowly lose some of their energy.

As mentioned in Chapter 2, William Henry Bragg had shown that the absorption of alpha particles was different from the standard absorption of light and x-rays. In the latter case, the amount of light decreases exponentially with depth. Alpha particles on the other hand have a definite range or so-called stopping power. The alpha particles slowly lose energy as they travel into a material and come to rest at a depth that depends on the initial energy of the particle and on the square root of the atomic weight of the matter. With modern computers, the paths of individual particles as they interact with both the nucleus and the electrons can be calculated. By summing the calculated trajectories of thousands of particles, the average depth is calculated. The calculations compare well with the original observations of William Henry Bragg. More importantly, the calculations are used to predict the distribution of dopants in silicon as a function of the energy of the dopant ions and the angle at which they hit the silicon crystal, so as to guide the manufacture of semiconductor devices.

The research done at the beginning of the twentieth century on the structure of the atom provided the foundation for developing ion implantation. By the time that Shockley proposed using accelerators to implant boron and phosphorous atoms in silicon to form transistors, the interaction of charged particles with matter was well understood, and ion implantation became one of the corner stones of the electronics industry.

5
The Sun Shines Bright

"Vast power of the sun is tapped by battery using sand ingredient" exclaimed a headline in the *New York Times* on 26 April 1954. The accompanying article announced the invention at Bell Labs of the first efficient solar cell based on a *p–n* junction in silicon. The sand ingredient, of course, referred to silicon. This chapter is about solar power and radiation detectors and the science behind them.

This was a propitious time for such an invention because the exploration of space was about to begin, and a renewable power source would be needed. Sputnik was launched just three years later in 1957, and the space race was on. Sputnik was powered by conventional batteries, and it stopped transmitting signals back to Earth when the batteries died. The Vanguard I satellite was launched in 1958, and it had both a battery powered and a solar powered transmitter. The latter continued to transmit signals for eight years until it failed because of radiation damage to the solar cells. TELSTAR, the world's first communications satellite, was built by Bell Labs and was launched by NASA in 1962. It was almost completely covered by silicon solar cells (Fig. 5.1).

In the ensuing half century solar powered satellites have become an integral part of our lives. Thousands of solar powered satellites have been launched. They are used as links in global telecommunications and information technology. There are satellite cell phones, satellite TV, weather satellites, GPS, and on and on. The exploration of the Moon and the planets are all dependent on solar power. The chance discovery by Russell Ohl of the *p–n* junction, and his observation that it converted solar radiation into a direct current—the photovoltaic effect—has blossomed into one of the enabling technologies that make space science possible.

Although satellites rely on harnessing solar energy, the terrestrial use of the Sun's energy comprises only a small fraction (1%) of our current energy use. Fossil fuels are used to satisfy most of the world's energy needs because, at present, they are plentiful and inexpensive. However, they are a finite resource and a source of pollution.

Attempts were made after the oil embargo in the 1970s to nurture the renewable energy industry, but the world lost interest as memories

Figure 5.1 TELSTAR communications satellite powered by 3600 Bell solar cells was built by Bell Telephone Laboratories and launched by NASA in 1962. (Reprinted with permission of Alcatel-Lucent USA Inc.)

of the embargo faded. With renewed concerns about the stability and price of oil supplies, and with increasing concerns about global warming, the world is trying again to develop renewable energy sources. In the long run, unlike fossil fuels, solar energy is the only renewable form of energy available to us. The energy delivered by the Sun to the Earth is 120,000 terawatt-years (TWyr). By comparison, human activity on earth uses about 14TWyr. In other words, the Sun delivers the equivalent of our annual energy needs every hour. The total known oil reserves are equivalent to about 1.5 days of solar energy.

The energy from the Sun can be harnessed in a number of ways. These include photovoltaic (PV) solar cells, direct solar heating, wind energy that results from the Sun heating the Earth and the atmosphere, or growing biomass that is subsequently converted to fuel. The challenge is to reduce the cost of solar power to make it competitive with fossil fuels and to store it for use as needed.

The cost of electricity from photovoltaic systems is presently between $0.25 and $0.65 per kilowatt hour, whereas the cost of electricity from coal is $0.04 per kWh (Slaoui and Collins 2007). Even with this cost disadvantage, solar power is competitive in remote areas where the higher initial cost of installation is offset by the lower maintenance and operating costs. Solar power installations are often competitive when an electric utility needs to build a new power plant. A utility company has to consider fluctuations in fuel prices, the potential future cost of carbon emissions, and the availability of water, when deciding between building a PV plant or a conventional fossil fuel power plant.

In addition, they have to consider that a PV plant can be constructed in a couple of years whereas a fossil fuel plant is constructed in 10–15 years (Swanson 2009).

Electric utilities around the country are installing PV plants. Figure 5.2 shows an 8 MW plant built for Xcel Energy, Inc. by SunEdison LLC. Worldwide production of photovoltaic devices was around 4 GW in 2007 and around 7 GW in 2008—a 60% increase in one year. (For a listing of the 50 largest PV plants worldwide see Lenardic (2011).) The largest plant as of the end of 2010 is a 97 MW plant built by First Solar, Inc. at Sarnia, Ontario Canada.

There are over a hundred companies in the US making PV systems, but the US produced only 7% of the world PV production in 2008, whereas Germany produced 57%. More and more governments are subsidizing their renewable energy industries to get them to where they are economically competitive, much as the US government did with the nascent semiconductor industry in the 1950s. Renewable energy will be an important industry in the twenty-first century, and the US needs to compete for market share and the accompanying job creation. Hardly a week goes by without an article appearing in the newspaper or on television about a new renewable energy project. It is encouraging that the world is focused again on making solar power a viable alternative to fossil fuels (Swanson 2006).

However, the PV industry faces enormous challenges in making large inroads in the energy business. The cost per kilowatt hour has to

Figure 5.2 8 MW solar power system installed in Alamosa, Colorado by SunEdison LLC. (Courtesy of Lawrence Kazmerski, National Renewable Energy Laboratory.)

decrease by a factor of 5 through better solar cells, reduced component costs, and economies of scale in manufacturing. For the last 35 years the cost of a PV module per watt of power has followed the standard industrial learning curve, where there is a 20% reduction in the module cost for every doubling of cumulative production. The module cost has decreased by a factor of almost 25 since the mid-1970s. These costs should continue to follow this learning curve especially with the present exponential growth in production. At the same time, R&D is leading to more efficient and cheaper solar cells, and more start-up companies are testing new ideas for photovoltaic modules. Technology roadmaps predict that solar power will become fully competitive with conventional power generation within a decade (Swanson 2009).

How is the energy of sunlight turned into an electrical current flowing out of a solar cell? The first indication that light can be converted directly to electricity dates back to experiments by Alexandre Becquerel in 1839. He found that when certain electrodes were immersed in an electrolyte and exposed to light a small electrical current was observed to flow between the electrodes. In the late 1880s, Willoughby Smith and Charles Fritts made solar cells using selenium and platinum as electrodes, but these were very inefficient devices. (A timeline of the history of solar power is at the US Department of Energy website: http://www1.eere.energy.gov/solar/pdfs/solar_timeline.pdf.)

Between the time of the demonstration of the initial solar cell and the invention of the silicon solar cell, the classical physics of the nineteenth century was replaced by the quantum physics of the twentieth century. The concept of the quantum of energy and the particle/wave duality of light came out of theories developed to explain the emission of light from hot bodies such as the Sun and the emission of electrons from the surface of metals after the absorption of ultraviolet light. (These experiments are discussed in Box 5.1.) In turn, quantum theory led to an understanding of how sunlight is absorbed and converted to electricity in photovoltaic devices.

BOX 5.1 The Quantum of Energy

The spectrum of the radiation emitted by the sun is shown in Fig. 5.3. (The data are from http://rredc.nrel.gov/solar/spectra/am1.5/ASTMG173/ASTMG173.html.) The intensity of the radiation increases with increasing energy. The intensity reaches a maximum and then falls at higher energies. The general features of this spectrum are observed in the emission of light from a whole class of hot objects such as the stars, the Sun, a blast furnace, or even the human body.

These spectra were studied during the last two decades of the nineteenth century, and a general theory to describe the features was developed in the first decade of the twentieth century. Experimental physicists try to understand nature by designing idealized systems and then making measurements on these systems. Then, they try to develop a theory that gives a mathematical description of the observations. For thermal radiation, the idealized system is called black body radiation.

Consider a hypothetical oven that is held at a constant temperature. There will be electromagnetic radiation composed of many different energies inside the oven. For a black body, it is assumed that any radiation that hits the walls is completely absorbed, that is it is "black." In turn, the walls continuously emit radiation so that over time an equilibrium is reached in which the amount absorbed by the walls equals the amount emitted. Black body radiation is the radiation that comes out of a small hole put in the wall of the oven after it has come to equilibrium.

One can study the intensity of black body radiation as a function of energy as the temperature of the oven is changed. It was shown that the total energy radiated per unit time varies as the fourth power of the temperature (Stefan–Boltzmann Law). Also, the energy at which the maximum amount of energy is emitted varies as the temperature (Wien Displacement Law). When these laws were derived theoretically, it was shown that the constants of proportionality in both laws were combinations of fundamental constants of physics. The shape of the curve of radiation versus energy and the energy of maximum radiated energy only depend on the temperature. A number of different objects show an intensity of thermal radiation and a spectrum that is close to that calculated for black body radiation at the appropriate temperature. Examples include the sun at T = 5800 °K (Fig. 5.3) and the human body at T = 300 °K.

To illustrate the importance of black body radiation to the development of twentieth century physics, two theoretical curves for the energy dependence of black body radiation are shown in Fig. 5.3. The first is the Rayleigh Jeans Law and the second is Planck's Law. The two curves start together but diverge as the energy increases. Planck's Law for a black body at T = 5800 °K agrees very well with the observed spectrum from the Sun. The important point is that the Rayleigh Jeans Law was derived using the laws of classical physics that had been developed in the nineteenth century. (Historically, the Rayleigh Jeans law was derived later than Planck's Law, but it is a useful pedagogical tool to illustrate the breakdown of classical physics.) The fact that this law fails to show a decrease in intensity of thermal radiation at high energies means that something is missing in the classical picture of thermal energy.

The explanation of this discrepancy ushered in quantum theory, and it dramatically changed our view of the world. The key insight was made by Max Planck in 1900. The fundamental difference between the Raleigh Jeans Law and Planck's Law lies in how one understands energy. Light is a form of energy, and it propagates as a wave. In the nineteenth century it was assumed that light was a continuous stream of energy and that the stream could be subdivided into arbitrarily small units. Since the time of the Greek philosopher, Democritus, matter was considered to be granular and made up of indivisible units that we now call atoms. So why wasn't energy also granular? In the seventeenth century, Newton had in fact proposed that light was made up of little packets of energy, but the

(cont.)

BOX 5.1 *Continued*

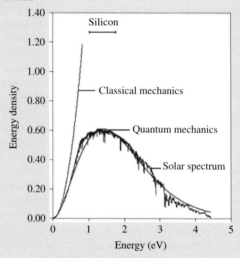

Figure 5.3 Planck's Law for T = 5800 °K agrees well with the solar spectrum. It is based on the quantization of light. The Rayleigh Jeans relation that is based on nineteenth century classical physics only agrees at low energies. Silicon made from sand is used to make the majority of solar cells because as indicated it absorbs light near the peak of the solar spectrum.

wave theory of light dominated nineteenth century thinking because it successfully accounted for virtually all experiments done using light. Planck returned to Newton's idea.

In black body radiation, the walls of the oven are continuously absorbing and emitting light. In the classical theory of light, arbitrarily small amounts of energy can be emitted or absorbed, and this assumption leads to the Raleigh Jeans Law that does not agree with the observations.

Planck developed a hypothetical model for the walls of the oven in which there are a number of oscillators, and each has its own frequency. In this model the energy contained in each oscillator at any given time is a multiple of a universal constant (now known as Planck's constant) multiplied by its frequency. When an oscillator exchanges energy with the light inside the oven by emitting or absorbing light, it does so in units of energy equal to Planck's constant multiplied by the frequency of the oscillator. Planck suggested that energy is in fact granular and that there is a minimum unit of energy, the quantum. The change from a continuous to a granular view of light fundamentally changes the mathematics of the calculation of thermal radiation as a function of energy. This leads to an equation that agrees with the measured solar spectrum as shown in Fig. 5.3. Planck received the Nobel Prize in 1918 "in recognition of the services he rendered to the advancement of Physics by his discovery of energy quanta."

The quantization in Planck's model for black body radiation comes from the hypothetical oscillators in the walls of the oven. Einstein had the deeper insight that it is the electromagnetic field of light itself that is quantized.

> Einstein had an outstanding year in 1905. He published four seminal papers that changed the course of physics. People associate Einstein with $E = mc^2$ (the equivalence of matter and energy) and with the theory of relativity, but in fact he received the Nobel Prize in 1921 "for his services to theoretical physics, and especially for his discovery of the law of the photoelectric effect." Whereas Planck had explained the emission of light from a hot body, Einstein explained how light of sufficient energy is absorbed by a metal and emits electrons. This is called the photoelectric effect.
>
> As with black body radiation, the photoelectric effect cannot be explained by classical physics. All metals emit electrons when irradiated with light of sufficient energy, but no electrons are emitted below a critical energy. Furthermore the critical energy is different for each metal.
>
> What is most surprising is that as the intensity of light with a given frequency increases, the number of emitted electrons increases, but the energy of each emitted electron doesn't change. This is counter intuitive if one thinks of light as a simple wave. When we watch the wind drive the waves in the ocean as a storm approaches, the amplitude of the waves increases as the strength of the wind increases. The energy is proportional to the square of the amplitude, and therefore the energy in the wave increases.
>
> In the photoelectric effect as more light of a given energy is absorbed, the energy of the emitted electrons doesn't change. Following Planck, Einstein realized that if light is granular then a particle of light (now called a photon) gives up all its energy to the electron. As the intensity of the light increases more electrons absorb photons and are emitted from the surface of the metal, but each has the same energy.
>
> A photon of a minimum energy is needed to cause an electron to break free of the metal. As the energy of the photons is increased above the critical energy, the energy of the emitted electrons increases because there is energy left over above the critical energy needed to free the electron. Einstein's theory of the photoelectric effect explains all of the experimental observations—the critical energy for the emission of electrons and the intensity of emitted electrons with increasing energy.
>
> Planck's quantum of energy and Einstein's photon of light coupled with quantum mechanics leads to an understanding of the absorption and emission of light in semiconductors, and in particular, in p–n junctions. This understanding in turn led to the development of solar cells and radiation detectors—products that have become integral parts of our daily lives.

Implicit in the discussion of semiconductors, p–n junctions, and transistors is the concept of the quantum of energy, the photon. In a solar cell, a photon is absorbed by an electron, and the electron is promoted from the valence band to the conduction band. This leaves a hole in the valence band. An electron-hole pair is formed in which there is an extra electron in the conduction band and an empty hole in the valence band. Over time, the electron can either fall back into an empty hole, thereby re-emitting a photon, or the electron and hole can each drift toward the appropriate electrode, producing an electric current.

Silicon can absorb photons with a range of energies. The valence and conduction bands are each spread out in energy, so an electron with energy between the bottom and the top of the valence band can be excited to an energy between the bottom and the top of the conduction band. It is fortunate that the range of photon energies that can be absorbed by silicon falls within the range of photon energies that are emitted by the Sun (Fig. 5.3). This is what makes silicon an attractive material to use in solar cells.

When sunlight falls on a piece of pure silicon, some of the light will be absorbed creating electron-hole pairs. In order to produce an electrical current, electrons and holes must drift toward the appropriate electrodes before they recombine. The p–n junction, discussed in Chapter 4, is the key. Because of the voltage barrier at the p–n junction, when an electron-hole pair is formed, the electron will drift toward the n-doped side and the hole will drift in the opposite direction toward the p-doped side. When a resistance is connected to each side of the p–n junction, electrons flow from the n-doped Si through the resistance and recombines with a hole in the p-doped Si. This is the essence of the p–n junction solar cell (Fig. 5.4).

Although this effect was discovered by Russell Ohl in 1941, manufacturing practical solar cells had to wait until both the position and the sharpness of the p–n junctions and the purity of silicon could be accurately controlled. The p–n junction was discovered when phosphorous and boron impurities in germanium separated from each other as the germanium solidified (Fig. 4.2). With the invention of the junction transistor in 1941, attention turned to the formation of p–n junctions. In the beginning, the junctions were grown from the melt. A crystal was grown by the Czochralski technique in which a seed crystal is slowly withdrawn from liquid silicon (Fig. 3.3). Pellets of either a boron or a phosphorous dopant were added to the melt, and the resulting crystal had a boundary between regions containing the different dopants. To make a device, a plate had to be cut out of the crystal just as Russell Ohl had done. A better way had to be found to produce p–n junctions.

Calvin Fuller and Gerald Pearson demonstrated that large area p–n junctions could be made by diffusing the dopant into the surface of a germanium or silicon crystal. To diffuse a dopant into a crystal, a compound of the impurity is passed over the surface of the crystal that is held at a high temperature to increase diffusion. The profile of the dopant in the surface region is determined by the temperature of the crystal and the time during which the dopant containing gas is flowing over the crystal. By the late 1940s, silicon solar cells were being made by diffusing phosphorous into p-silicon to form an n–p junction near the surface, or by diffusing boron into n-silicon to form a p–n junction (Millman 1983, p. 431). A solar cell with an efficiency of 6% was achieved in 1954 by Daryl Chapin, Calvin Fuller, and Gerald Pearson at Bell Labs, and they filed a patent on a "Solar Energy Converting Apparatus." This patent was the basis for the headline in the *New York Times* (Chapin et al. 1954). The research carried out during the ensuing half century has concentrated on trying to increase the overall efficiency of solar cells and to reduce manufacturing costs.

One of the most efficient silicon solar cells developed by Martin Green and his colleagues in New South Wales is shown in Fig. 5.4. It has an efficiency of 24%—four times higher than the cell designed by Chapin, Fuller, and Pearson. This efficiency approaches the theoretical limit for a single junction cell of 32% (Shockley and Queisser 1961).

This cell illustrates a number of the improvements in cell design that have been made over the past half century. Collecting all of the light that falls onto the solar cell is not as simple as it might seem. Some passes through the cell without being absorbed. Some is reflected back. And, if the cell is too thick, the electrons and holes may recombine before reaching the electrical contacts. So a balance has to be achieved.

One idea that is illustrated in the figure is to texture the surface of the crystal so the light bounces back and forth inside the crystal, increasing the probability of absorbing the photon. When a crystal is put in an appropriate acid, different crystallographic planes will be dissolved or etched at different rates. When a plate of silicon is etched, the resulting surface of the crystal is composed of a network of inverted pyramid-shaped depressions (voids) formed because some planes of silicon etch faster than others. (Specifically the d_1 planes in Fig. 2.2 etch faster than the d_4 planes that are parallel to the surface of the silicon plate.) Sunlight entering through this textured surface is reflected multiple times.

A thin film of an anti-reflective coating is deposited on the cell to reduce the probability of a photon escaping from the crystal. Anti-reflective coatings include our ubiquitous film of silicon dioxide or tin oxide. Another way to increase efficiency is to balance the production of electron-hole pairs with their recombination rates. The mobilities of electrons and holes in silicon are different. To balance the time it takes

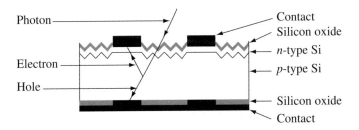

Figure 5.4 High efficiency (24%) silicon solar cell (adapted from Green 2002). A photon enters through a textured surface that along with the antireflective coating increases the probability that the photon will be absorbed forming an electron-hole pair. The electron diffuses to the contact in the n-type Si and the hole diffuses to the contact in the p-type Si producing a voltage between the contacts.

for electrons and holes to reach the electrical contacts, the *n*-doped region at the top of the cell is made thinner than the *p*-doped region.

To collect more of the Sun's light, solar cells are composed of stacks of *p–n* junctions. Each *p–n* junction is made from a different semiconductor, each with a different band gap so as to collect a different part of the Sun's spectrum. In theory, such a layered solar cell could achieve efficiencies of 66%. An example of a multi-junction solar cell is shown in Fig. 5.5. It is composed of pure amorphous silicon and various amorphous silicon–germanium alloys. Each layer is *p*- or *n*-doped at the top and bottom respectively so as to give a sandwich of three solar cells stacked one on top of the other. The bandgap in amorphous silicon is 1.7 eV, which is substantially larger than the gap in single crystal silicon (1.1 eV). The bandgap in each cell is changed by varying the composition of the silicon–germanium alloy so as to absorb a larger fraction of the Sun's spectrum.

A number of different semiconductor materials can be used to make multi-junction solar cells (CdS/CdTe and CdS/InP among others). In these complicated devices, layers of atoms are grown one layer at a time. Controlling growth at the level of one atomic layer is another great achievement of twentieth century science, discussed in Chapter 6.

The evolving field of nanotechnology has yielded yet another way to control the band gap in a semiconductor. The band gap is found to vary with the size of the semiconductor particle. By making arrays of semiconductor dots of differing size, the absorption of light across a majority of the solar spectrum can be achieved.

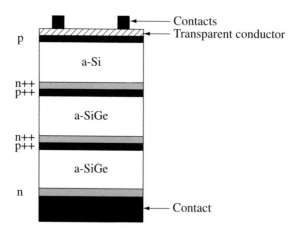

Figure 5.5 Thin film triple junction solar cell made of amorphous silicon and amorphous silicon germanium alloys. The alloy compositions are adjusted so each cell has a different energy gap. The three cells cover a larger fraction of the solar spectrum and increase the overall efficiency of the solar cell.

The limiting factor in the widespread utilization of photovoltaics is the cost of making and processing silicon or other semiconductors. The cost of manufacturing large solar cell arrays benefits from the manufacturing sophistication developed by the electronics industry over the past half century. Costs can also be lowered by using less semiconductor material. As a result, current solar cells are only 100 micrometers thick.

Another avenue is to make solar cells from thin films of microcrystalline or amorphous silicon instead of large expensive single crystals, as illustrated in Fig. 5.5. They are cheaper to produce but have lower efficiencies. The market share of these cells is increasing, but at the present time single crystal p–n junction silicon solar cells still account for 85% of the market for photovoltaics (Crabtree and Lewis 2007).

There are steady improvements being made to lower the cost and increase the manufacturing efficiency of solar cells. With each improvement the industry becomes more competitive compared to fossil fuels. There is every reason to believe that innovative scientists and engineers will succeed in making renewable energy an economic reality in the coming decades.

In a solar cell, radiation in the form of light is absorbed and produces a voltage across the cell. The solar cell is a detector of radiation. If the sun shines, there is a voltage, and the output of a solar cell is a good quantitative measure of the intensity of the radiation. Many different types of radiation detectors are based on p–n junctions in silicon. These include photodiodes for optical communications, CCD (charge coupled devices) arrays for digital cameras, and particle detectors for high energy physics experiments.

In optical communications (Chapter 7) a transmitter converts an electrical signal into a light signal that in turn travels through an optical fiber. At the other end of the fiber there is a photodiode to convert the light signal back to an electrical signal. For the band of optical frequencies around 850 nm, silicon photodiodes are widely used.

The operation of a photodiode is different from a solar cell. In a solar cell, the electrons and holes resulting from the absorption of light are driven away from the junction and toward the electrodes, thereby producing a photocurrent. In the absence of light, there is a balance between the electron-hole pairs that are produced thermally and the recombination of existing electron-hole pairs. If a reverse bias voltage (p-side is negative and n-side positive) is applied, then recombination is greatly reduced because the electrons and holes are driven away from the p–n junction. This mode of operation greatly increases the sensitivity of the photodiode when it is used as a detector.

The efficiency of a photodiode is further improved by increasing the width of the depletion region. The width can be increased by a layer of intrinsic (undoped) silicon inserted between the n and p type regions as illustrated in Fig. 5.6. (This is similar to the multi-junction solar cell

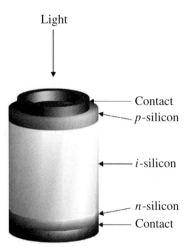

Figure 5.6 A p–i–n diode detector. Light enters through the top, and it is absorbed in the i-silicon forming electron-hole pairs. Under reverse bias electrons flow through the n-silicon to the lower contact, and holes flow through the p-silicon to the top contact. This results in a change in the current flowing through the p–i–n diode that is proportional to the intensity of the incident radiation. p–i–n diodes made of silicon and indium gallium arsenide are essential components of optical communications systems.

shown in Fig. 5.5.) This layer increases the volume of the region where electron-hole pairs can be formed. These detectors are called p–i–n diodes, which stands for a device with positive–intrinsic–negative regions. *Pin* diodes, or a more sophisticated version called avalanche photodiodes, are the receivers in most optical communications systems.

Silicon-based CCD detectors are used to record images in cell phones, digital cameras, and camcorders. These images are composed of arrays of pixels, and digital cameras are advertised as having 4, 8, 10, and so on megapixels. Each pixel starts with a tiny photodiode constructed to store an electric charge proportional to the number of electron-hole pairs generated by the light falling on the pixel.

To store a digital image, the charges on all of the pixels have to be converted from a two-dimensional array to a linear sequence of voltage pulses. The amplitude of each pulse is proportional to the charge stored in a pixel. Each voltage pulse is converted to a digital signal in an analog-to-digital converter. The disk in the camera stores the sequence of digital signals.

A charge coupled device (CCD) is used to transfer the charge of each pixel to a shift register. The shift register is a sequence of capacitors, where the bias voltager is alternatively turned off and on to transfer

THE SUN SHINES BRIGHT | 99

Figure 5.7 Charge Coupled Device (CCD) for the detection of visible and ultraviolet light. In addition to use in consumer products like digital cameras, CCD sensors are an enabling technology for space based astronomy like the Hubble Space Telescope. (Courtesy NASA/JPL-Caltech.)

Figure 5.8 Two hundred silicon particle detectors arranged in three concentric barrels are a part of the STAR detector at the Relativistic Heavy Ion Collider. (Courtesy of Brookhaven National Laboratory.)

the charge. This process is used to convert the charges stored on each pixel in a two-dimensional array into a linear sequence of charges in a register. The CCD was invented in 1969 by Williard Boyle and George E. Smith at Bell Labs, and they shared the Nobel Prize along with Charles Kao in 2009.

Detectors of ionizing radiation like x-rays are also based on p–i–n diodes. The ionizing radiation is absorbed in the diode producing a number of electron-hole pairs that is proportional to the energy of the x-ray.

Arrays of silicon detectors are also used as particle detectors in high energy and nuclear physics experiments at the LHC (Large Hadron Collider) at CERN, the TEVATRON at Fermi Lab and RHIC (Relativistic Heavy Ion Collider) at Brookhaven National Laboratory. The STAR detector at RHIC has over 200 silicon drift detectors arranged in three concentric barrels (Fig. 5.8). The banks of silicon detectors analyze the energy and direction of the charged particles that result when high energy beams collide with each other. The direction of the particle is determined by the position of the detectors with respect to the interaction region where the counter-circulating particle beams collide.

Solar cells and detectors evolved from grains of sand to pure silicon to p–n junctions. The combination of progress in physics over a hundred years and of progress in making new materials has led to reliable sources of electrical power and to detectors for optical communications, digital photography, and space based astronomy.

6
How Small is Small?

Information technology is a central part of our lives. We expect an array of new consumer electronics every year. The gadgets keep getting smaller, and they have more features. Over four billion people have cell phones. This progress depends on making electronic components smaller, faster, and cheaper. In order to make smaller and more sophisticated components, materials are made and controlled on the atomic scale. Crystals are grown atomic layer by atomic layer.

Entering the world of layer by layer growth requires understanding of the properties of surfaces and the structure of surfaces because they are different from the properties and structure of the bulk or interior of crystals. To do this, twentieth-century science developed techniques to watch a crystal of one material grow on the surface of a different material and to determine the structure of the surfaces of crystals. The architecture of sand is elucidated using x-ray diffraction, and this technique grew out of experiments to demonstrate the wave nature of x-rays. Similarly, the structure of the surface of a crystal is studied using electron diffraction, and this technique grew out of experiments that demonstrated the wave nature of electrons (Box 6.1). Out of these experiments grew low energy electron diffraction (LEED) that is used to determine the structure of the surfaces of crystals. An extension of this technique, reflection high energy electron diffraction (RHEED), is used to monitor the growth of a single crystal—atomic layer by atomic layer in real time.

> **BOX 6.1 Wave Nature of the Electron**
>
> In 1897, J. J. Thomson demonstrated that electrons, which were then called cathode rays, were particles, and he determined the mass and the charge of the electron. For this discovery he was awarded the Nobel Prize in 1906. During the 1920s, Clinton Davisson and his collaborators at Bell Labs were trying to improve the performance of vacuum tubes. They started a basic research program on the interaction of beams of electrons with the surface of nickel. They had a fortunate accident, as it turned out, when their vacuum chamber broke exposing the nickel
>
> (cont.)

BOX 6.1 *Continued*

sample to the air. As the sample was at a high temperature, its surface was heavily oxidized. In addition to repairing the vacuum system, they had to clean the surface of the nickel sample by heating it in vacuum to drive off the oxygen. When they returned to their experiments with beams of electrons, they found that the intensity of the scattered electrons exhibited spikes as the orientation of the nickel was changed.

When the Ni was heated to high temperature in a vacuum, a number of fairly large single crystals had grown in the sample. Davisson realized, by analogy with x-ray diffraction, that he was observing the Bragg scattering of electrons from the crystals. As discussed in Chapter 2, Bragg's Law relates the wavelength of the radiation (λ) to the spacing between planes of atoms in a crystal (d): $n\lambda = 2d\sin(\theta)$ where 2θ is the angle of the scattered electrons relative to the incident beam of electrons. The spikes in intensity occurred when one of the crystals had the correct angle to satisfy Bragg's Law. This result was entirely unexpected. X-rays propagated as waves, but electrons were particles, and they shouldn't exhibit wavelike behavior.

However in 1926, Louis deBroglie had proposed in his doctoral thesis that in quantum mechanics, particles did have wavelike properties. He predicted that the wavelength of a particle (λ) in this case—electrons—was related to the momentum of the particle (p) by Planck's constant (h): $\lambda = h/p$. This proposal was revolutionary.

Davisson's experiment provided a direct proof of the deBroglie hypothesis. Davisson could measure both the wavelength and the momentum of the electron independently to see if they were related by Planck's constant. The momentum is calculated from the velocity because momentum is the product of mass multiplied by velocity. The electron mass had been determined by J. J. Thomson, and the velocity is determined by the voltage with which the electrons are accelerated. The wavelength is calculated from Bragg's Law using the measured scattering angles and the spacings between layers of atoms calculated from the crystal structure of nickel. The structure had been determined previously using x-ray diffraction. This wavelength calculated from Bragg's Law was then compared with the wavelength calculated from the deBroglie relation, and they agreed. The wave/particle duality of the quantum world had been established (Davisson and Germer 1927).

Quite independently, George Thomson, the son of J. J. Thomson, working at the University of Aberdeen in Scotland, also observed the diffraction of electrons. In his experiment, a beam of high energy electrons passed through a thin crystal, and the diffraction pattern was recorded on a fluorescent screen. (This geometry is similar to that used by von Laue, Fredrick, and Knipping to observe the diffraction of x-rays.) deBroglie was awarded the Nobel Prize in 1929 "for his discovery of the wave nature of electrons," and Davisson and Thomson were awarded the Nobel Prize in 1937 "for their discovery of the diffraction of electrons by crystals."

Unlike other probes of the structure of materials such as x-rays and neutrons, electrons are charged particles, and interact strongly with the atoms in a crystal. Low energy electrons are absorbed or scattered by the atoms at or near the surface of the crystal, and therefore, the diffraction of electrons is a sensitive probe of the arrangement of the atoms at the surface. Over the next 40 years, low energy electron diffraction (LEED) grew to be essential to understand how the structure of surfaces changes with temperature or with exposure to various gases.

Reflection high energy electron diffraction (RHEED) is used to monitor the layer-by-layer growth of crystals. A beam of high-energy electrons (40 keV) diffracts from the surface and onto a fluorescent screen. The deBroglie wavelength of the electrons is 0.006 nm, about an order of magnitude smaller than the height of an atomic step on the surface of a crystal of a typical semiconductor like Si (0.27 nm).

It is useful to think about the optical analog of the reflection of light from a mirror. If the mirror is perfectly smooth, the mirror will strongly reflect the light. If the mirror surface is rough, the reflected light will be weaker. In the case of the growth of a monolayer of atoms on the surface of a crystal, the surface is atomically smooth before and after the monolayer is deposited. In between, there will be many terraces on the surface, and the surface is rough. As a result, the intensity of the diffracted electrons oscillates during the deposition of a monolayer. The intensity decreases as the terraces start to grow. It is a minimum when half of a monolayer has been deposited and the maximum number of terraces (roughness) is reached. There is a direct correlation between the period of the oscillation in the intensity of the scattered electrons and the deposition of one atomic layer. In this way, the deposition of the layers can be monitored during growth layer by layer.

The use of these techniques has led to advances such as faster computer chips, light emitting diodes (LEDs), semiconductor lasers, and multiple junction solar cells. At the same time, these new man-made crystals have new unanticipated properties, and the discovery of these new properties of matter has led to four Nobel Prizes.

For four decades, the increases in chip density have followed Moore's Law, which predicts that industry can double the number of transistors on a computer chip roughly every 18 months. A computer chip from Intel can contain over a billion transistors on a postage stamp sized single crystal of silicon. That number is equal to one sixth of the world's population. If the industry can continue to follow Moore's law for another six years or so, then the number of transistors on a single chip will equal the population of the world.

What are the ultimate limitations on the size of a field effect transistor? There are a number of challenges on the horizon, which become more daunting as devices get smaller. Materials scientists are searching for ways to improve performance that do not just rely on shrinking the dimensions of individual field effect transistors.

There are several intrinsic properties of silicon that control the speed (mobility) with which electrons and holes move through a crystal. One way to increase the mobility of electrons is to change the symmetry of the crystal structure of silicon in the active region of the field effect transistor. Lowering the symmetry of silicon increases the mobility in certain directions.

The structure of silicon is cubic and, as a result, the energy gap between the valence and the conduction band is the same along each of the three cubic directions. If the crystal is compressed in one direction or alternatively is stretched in two directions, then the energy gaps are no longer the same in each direction. The mobility of the electrons increases in the directions perpendicular to the axis of compression. If a way can be found to do this in the active area in a field effect transistor or a junction transistor, then the mobility and, consequently, the device performance can be increased.

Another factor that affects the mobility of electrons and holes is the presence of dopant atoms. These atoms occupy sites in the crystal structure that are normally occupied by silicon atoms. The dopant atoms have either a larger or smaller nuclear charge than a silicon atom depending on whether the active region is n- or p-doped. This results in local variations in the electrostatic potential that the electrons and holes experience as they move through the crystal and, in turn, affects their mobility. If the dopant atoms are incorporated into films grown on either side of the active region, then the dopants will still supply the extra electrons or holes to the active region that are needed for conduction, but the nuclei of the dopant atoms will not get in the way, thereby increasing the mobility. This technique is called modulation doping.

The reduction of the symmetry of the silicon in the active region of an FET is called strain engineering. One application of strain engineering is in the strained silicon-metal-oxide semiconductor field effect transistors that are incorporated into the latest INTEL microprocessors. This device is sketched in Fig. 6.1 (Mohta and Thompson 2005). The source and drain, which are composed of n-silicon in a normal FET (Fig. 4.3), has been replaced by a silicon–germanium alloy.

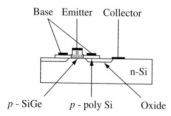
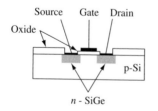

Figure 6.1 (top) Bipolar heterojunction transistor with a strained SiGe layer (Ashburn 2003). (bottom) High mobility field effect transistor. The source and drain are epitaxially grown SiGe alloys that induce a strain in the silicon channel below the gate. The SiGe alloys increase the mobility of the electrons or holes and the performance of both devices.

The silicon–germanium alloy has to be grown as an almost perfect crystal that is aligned with the underlying silicon. If there are too many defects, then the performance of the device will deteriorate. During manufacture, the silicon in those regions is removed by etching, and then crystals of a silicon–germanium alloy are grown in those regions.

The unit cell of a silicon–germanium alloy is larger than that of silicon, and as a result the silicon between the source and the drain is compressed. This strain increases the mobility of the electrons. By building on the control of the layer by layer growth of silicon–germanium, the strain and composition can be tailored to achieve greatly improved performance of FETs without shrinking the size of the components.

One can only marvel at the degree of sophistication in this process. First is the ability to grow a single crystal of one type of material, in this case a silicon–germanium alloy, imbedded inside a single crystal of another material, silicon. Second is the realization that this technology was first introduced when the distance between the source and the drain had been reduced to 90 nm. That distance is a thousand times smaller than the diameter of a human hair. Also, one must not forget that this process is repeated in millions of FETs on the same chip—a remarkable achievement in manufacturing.

In addition to strained field effect transistors (FETs), there are strained bipolar junction transistors that are used in cell phones. In the latter the emitter, base, and collector are made respectively from n-type, p-type, and n-type silicon to form an n–p–n junction transistor (or p–n–p junction transistor). The performance of an n–p–n junction transistor is improved if the p-type region is replaced with a thin layer of SiGe, as shown schematically at the top of Fig. 6.1. The n–p and p–n junctions made between different materials are called heterostructures or heterojunctions. The resulting transistor is referred to as a heterostructure bipolar transistor, or HBT.

As in the FET above, a silicon–germanium film is grown epitaxially between silicon layers so that it is strained, improving the mobility of the electrons. In addition, the Ge composition is changed across the active region from the emitter to the collector. As a result, the energy of the bottom of the conduction band changes, and this drives the minority carriers through the base region. All of these effects increase the maximum frequency at which the device can operate.

The rapid expansion of the cell phone and other wireless industries has led to an increase in the performance of HBTs, and operation at higher and higher frequencies. Competition is so strong that the first successful HBTs were demonstrated by three different groups in the same month, December 1987 (Patton *et al.* 1988; Temkin *et al.* 1988; and Xu *et al.* 1988). Since then, there have been several generations of HBTs with decreasing thickness of the SiGe film (0.25, 0.18, and 0.12 microns) and correspondingly higher maximum cutoff frequencies

(50, 120, and 210 GHz). These devices find applications in both the handsets and base stations of cell phones and in a number of other wireless systems (Cressler and Niu 2003).

These devices are a result of the blossoming of surface science that occurred during the last quarter of the twentieth century. Different surface sensitive techniques were developed, which has led to an understanding of how to grow materials atomic layer by atomic layer. There is now an alphabet soup of techniques to grow single crystal films of one material epitaxially on a single crystal of another material. (An epitaxial film has the same orientation as the underlying crystal or substrate, and it is derived from the Greek words "epi" for above and "taxi" for ordered.) These techniques go by names such as liquid phase epitaxy (LPE), vapor phase epitaxy (VPE), metal-organic vapor phase epitaxy (MOVPE), and molecular beam epitaxy (MBE), among others.

Thin film growth techniques build on decades of experience growing crystals. As a crystal grows, atoms attach to the surface and then diffuse around on the surface until they reach an equilibrium lattice position. Many of us have grown crystals from solution as part of science experiments at home or in school. A solution of a salt in water is allowed to evaporate, thereby increasing the concentration of the salt. At some point, the system separates into two phases: a liquid phase (the solution) and a solid phase composed of crystals.

The number and size of the crystals that form depend on the balance between the rate at which microcrystallites form and the rate at which individual crystals grow. If the evaporation is too rapid, then many crystallites form and grow together producing a polycrystalline material. If the rate is slower, then only a few crystallites form, and each of them grows into a larger crystal.

The gems and minerals that are found in rock formations are grown on geological time scales out of saturated solutions, and the conditions favor the growth of large crystals. In the laboratory, the number of crystals that are grown is controlled by introducing a seed crystal. The growth rate is controlled by varying the flux of atoms arriving at the surface of the seed crystal. An example is the hydrothermal growth of large synthetic crystals of quartz that was discussed in Chapter 1. In the commercial production of silicon, a large perfect crystal is grown by slowly withdrawing a seed crystal from a container of liquid silicon as discussed in Chapter 3.

In the growth of a single crystal film of a silicon–germanium alloy in the FETs and HBTs discussed above, the seed is the underlying single crystal of silicon. It has been patterned by lithography to expose the areas where the silicon–germanium alloy is to be grown. Beams of silicon and germanium atoms (or compounds of silicon and germanium) are directed at the exposed areas, and the film slowly grows.

The success in growing strained layers of silicon–germanium on silicon is the result of decades of research. An early example of this research is the growth of films of the semiconductors gallium arsenide (GaAs) and aluminum arsenide (AlAs). This led to the development of light emitting diodes (LEDs) and semiconducting lasers that are pervasive in everything from lighting to barcode scanners and are key components in optical communications (Box 6.2).

The rest of this chapter reviews some of the research in layer-by-layer growth. We start with single crystals and then progress to the growth of crystals of one material on top of a different material. Finally, we discuss sandwiches (superlattices) composed of alternate regions of two different materials. The historical development centers on GaAs and AlAs, but the principles have been applied to silicon and germanium, as illustrated above in strained FETs and HBTs.

Consider the growth of GaAs using the technique of molecular beam epitaxy (MBE) that was developed by John Arthur and Al Cho at Bell Labs (Cho 1971). The electrical engineering, materials science, and physics departments at the major research universities all have MBE systems tailored to growing films of metals, semiconductors, or oxides. These are enormously complicated systems built inside stainless steel ultra-high vacuum chambers. At the center is usually a large spherical shaped chamber with a dozen flanges protruding from the surface of the chamber. Attached are various sources of gallium, arsenic, and other semiconductors used to grow the films. Then, there are devices that are used to clean the surface of the crystal on which single crystal films will be grown, devices to monitor the growth of the films, and devices to characterize the films.

The films have to be grown in an ultra-high vacuum in order to avoid contamination of the films, so there are several stages of pumps to reduce the pressure from one atmosphere to 10^{-10} atmospheres. The MBE chamber is usually wrapped with heating tapes, and these in turn are covered with aluminum foil. The chamber is baked to drive off gases that might be adsorbed on the inside walls of the chamber, and the aluminum foil reflects the heat back toward the chamber to increase efficiency. All of this gives the MBE apparatus the look of something out of a science fiction movie.

The apparatus has antechambers so samples can be transferred into and out of the chamber without affecting the ultra-high vacuum conditions inside the chamber. There are robotic systems inside the chamber to transfer samples around and shutters to open and close the beams of ions during growth, under computer control. Scientists are constantly monitoring all of the instruments to track progress and detect problems. It is a far cry from the simple glass vacuum systems used by Davisson and Germer to study the diffraction of electrons from the surface of nickel in the 1920s.

Achieving layer by layer growth by molecular beam epitaxy requires balancing a number of different effects. Beams of gallium and arsenic atoms are directed toward the clean surface of a crystal of gallium arsenide. The velocities of the atoms hitting the surface and the temperature of the gallium arsenide crystal have to be carefully adjusted, so that the atoms stick to the surface and diffuse around the surface to find an appropriate lattice site. If the temperature is too low, then the atoms stick where they land, and the resulting surface will not be atomically smooth. It will be composed of an array of hills and valleys. If the temperature is too high, then the atoms will evaporate from the surface.

One would like to be able to grow a film one monolayer at a time. As the layer starts to grow, small terraces form. If the temperature is just right, then as atoms are adsorbed on the surface, they diffuse across the surface until they come to rest at the steps that form the edges of the terraces. The terraces increase in size and a few of the impinging atoms start a second layer on top of an existing terrace or start new terraces. Eventually, a complete monolayer of GaAs has been deposited, and new terraces start to form a new monolayer on top of the one that has just finished growing. By balancing the velocity distribution of the molecular beams from the sources and the temperature of the substrate, a perfect single crystal film can be grown atomic layer by atomic layer. This process is referred to as homoepitaxy because a film of the same substance is being grown epitaxially on the substrate.

Can we grow a single crystal film of one material on top of a substrate composed of a single crystal of a different material—a process referred to as heteroepitaxy? For example, can we grow a single crystal of aluminum arsenide on a single crystal of gallium arsenide? The two materials have the same crystal structure, but the unit cells are not the same size.

The unit cell of AlAs (0.562 nm) is smaller than that of GaAs (0.565 nm). If layers of AlAs are placed on top of layers of GaAs, they wouldn't match up. As Al atoms bond to the surface of GaAs, the AlAs bonds stretch in order to match the GaAs bonds in the plane. In other words the AlAs film is strained so that the unit cells of the AlAs expand in two directions in the plane of the film and contract in the direction perpendicular to the film. This maintains a constant volume of the unit cell as illustrated on the top of Fig. 6.2.

The growth of the strained layers cannot continue indefinitely because it costs energy to strain the film to fit the substrate. The strain energy in the film increases as the film grows thicker. At some point, it is energetically favorable to lower the strain energy via the formation of misfit dislocations (van der Merwe 1962). An example of one way in which the strain can be relieved is shown on the bottom of Fig. 6.2 where the film has one additional diagonal row of AlAs compared to the GaAs substrate.

The perfect registry of the strained AlAs with the GaAs has been broken where a sheet of AlAs has been added. The average unit cell is

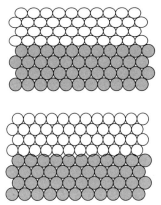

Figure 6.2 Heteroepitaxial growth of a crystal of one material on another material that is 10% larger (in real materials the difference is <1%). The smaller atoms on top are strained (shown as ellipses) so that their atomic spacing increases parallel to the plane to match that of the atoms in the bottom material. Their spacing perpendicular to the plane decreases to conserve the overall volume. In the lower figure, an extra diagonal row of atoms (misfit dislocation) relieves the strain. There is a critical thickness above which the strain is relieved by the formation of misfit dislocations.

therefore slightly smaller. As the film grows thicker, more and more dislocations appear until the AlAs film has fully relaxed to the unit cell size of pure AlAs.

There is a critical thickness above which misfit dislocations start to appear in a film. If the growth of the heteroepitaxial film is limited to thicknesses that are less than the critical thickness, then one has a single crystal that is composed of layers of GaAs followed by strained layers of AlAs, that is, a man-made composite single crystal.

Leo Esaki and R. Tsu at IBM suggested that multilayer structures composed of alternating regions of different semiconductors would lead to man-made materials in which the electronic properties are markedly different from normal semiconductors (Esaki and Tsu 1970). For example, a multilayer composed of four atomic layers of AlAs alternating with four layers of GaAs has an energy gap that varies periodically as illustrated in Fig. 6.3.

A multilayer is grown by alternately opening and closing the shutters that turn on and off the Al and Ga molecular beams. The timing of the

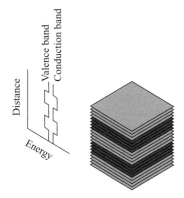

Figure 6.3 A multilayer composed of four layers of GaAs (grey) alternating with four layers of AlAs (black). On the left, the change in the energy of the valence band and the conduction band in the different layers is plotted in the direction perpendicular to the layers. The electrons in the GaAs layers are confined in the vertical direction because of the energy barriers at the interface with the AlAs layers.

shutters is coordinated with the RHEED oscillations (Box 6.1) to produce multilayers with sharp interfaces between GaAs and AlAs layers. An image of an $[(AlAs)_2(GaAs)_2]_n$ multilayer taken using transmission electron microscopy is shown in Fig. 6.4. (The chemical formula indicates that a sequence of two layers of AlAs followed by two layers of GaAs is repeated n times.) The multilayer was grown using molecular beam epitaxy by Art Gossard, and the image was obtained by Pierre Petroff when both of them were at Bell Labs. The image clearly shows the different layers. A scale is superimposed on the image, and it shows that the periodicity of the layers is close to that expected for alternating unit cells of AlAs and GaAs. X-ray and electron diffraction studies demonstrate that these multilayers are almost perfect single crystals that have grown epitaxially on a bulk GaAs single crystal. There is some intermixing of Ga and Al at the interface, but it is a remarkable achievement to grow a crystal atomic layer by atomic layer while changing the chemical composition with almost perfect periodicity (McWhan 1985).

The techniques to grow multilayers continue to improve, and multilayers composed of a wide variety of semiconductors, metals, and insulators are routinely grown atomic layer by atomic layer. This research in layer by layer growth led to the new strained silicon FETs and HBTs discussed above and the light emitting diodes and semiconductor lasers discussed in Box 6.2.

A fundamental property of a semiconductor is the energy gap. Different semiconductors have different energy gaps, so a multilayer

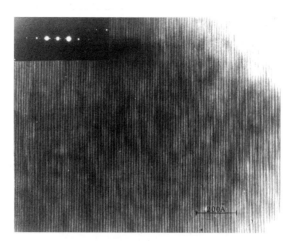

Figure 6.4 Transmission electron microscope image of a $[(GaAs)_2(AlAs)_2]_n$ multilayer crystal in which two layers of GaAs alternate with two layers of AlAs and the sequence is repeated n times. The inset is an electron diffraction pattern. The weak spots on each side of the strong spots result from diffraction by the superlattice. (Courtesy of Pierre Petroff, Univ. of California Santa Barbara.)

composed of different semiconductors has energy gaps that increase and decrease periodically. In the example above, the band gap in the multilayer alternates between 1.43 eV (GaAs) and 2.16 eV (AlAs) as illustrated in Fig. 6.3. Similarly a multilayer made of Si and Ge has energy gaps that alternate from 1.2 eV to 0.667 eV. (In most multilayers, in order to stay below the critical thickness for the formation of dislocations, alloys of GaAlAs and SiGe alternate with GaAs and Si respectively.)

A single crystal composed of alternating regions of different materials is called a heterostructure, and it exhibits new properties that are not observed in naturally occurring three dimensional materials. Electrons in the conduction band of GaAs are confined between the AlAs regions because the increase in energy at the interface between GaAs and AlAs acts as a barrier. If the energy of an electron is less than the barrier, then it is confined between the barriers. The situation is like a bowling alley where individual bowling balls (electrons) roll down a particular alley but are prevented from jumping to an adjacent alley by the gutters.

When the electrons are confined, the energies of the electrons are limited to discrete values because of quantum mechanics, and they are said to occupy quantum wells. A classical way to illustrate this quantum mechanical effect is to consider a vibrating string whose ends are attached to the barriers. In order to sustain a vibration, there has to be an integral number of half wavelengths between the barriers.

The confinement of electrons in a heterostructure was first demonstrated by Ray Dingle and co-workers at Bell Laboratories (Dingle *et al.* 1974). This is a layer structure so the electrons are only confined in the direction perpendicular to the barriers, and they are free to move within the plane. Being able to grow materials layer by layer provides a new way to control their properties through quantum mechanical effects.

The conductivity of a semiconductor is proportional to the number of dopants that are added. As mentioned above, the mobility of the electrons is limited in part by the fact that they scatter off the fixed dopant atoms. In 1978, Dingle and co-workers demonstrated modulation doping that is a clever way to greatly increase the mobility of the electrons (Dingle *et al.* 1978). Because of the exquisite control provided by MBE, they grew multilayers in which the dopants were deposited only in the regions on each side of the GaAs. The resulting electrons migrated to the quantum wells in the GaAs. They form a two-dimensional electron gas (2-DEG) in which the electrons can move freely in two directions but are confined in the third direction. As the scattering centers are no longer in the GaAs, the mobility of the electrons is increased.

Totally new phenomena have been observed in the two-dimensional electron gas that results from the quantum confinement of the electrons, and they have led to two Nobel Prizes. In 1985 Klaus von Klitzing won the prize for "the discovery of the quantized Hall Effect" and in 1998

Robert Laughlin, Horst Stormer, and Dan Tsui received the prize "for their discovery of a new form of quantum fluid with fractionally charged excitations."

In a 2-DEG, electrons flow freely in the GaAs layer under the influence of a voltage applied across the ends of the layer. If a magnetic field is applied parallel to the layer but perpendicular to the direction of the electron flow, then a voltage develops parallel to the magnetic field. This is called the Hall voltage after Edwin Hall who first observed the effect in 1879 when he was a graduate student at Johns Hopkins University. At very low temperatures and high magnetic fields, the electrons in a 2-DEG become quantized and can only have certain energies. Von Klitzing discovered that under these conditions, the Hall voltage exhibits discontinuous steps as the magnetic field is increased. The steps occur at integral multiples of a universal constant that equals the square of the charge on the electron divided by Planck's constant.

Horst Stormer and Dan Tsui at Bell Labs, using very pure heterostructures prepared by Art Gossard, found that at even lower temperatures and higher magnetic fields, new steps appeared in the Hall voltage. These occurred precisely at simple fractions of the same constant. Bob Laughlin, who was also at Bell Labs at the time, developed a theory that explained all of the steps. At very low temperatures and high magnetic fields the electrons are compelled to form a new collective state of matter called a quantum fluid. Examples of other quantum fluids are the superconducting state of some metals at low temperature and the superfluid state of liquid helium. These are all visible manifestations of the strange world of quantum mechanics.

A completely new unrelated phenomena was discovered in multilayers composed of alternating regions of different magnetic materials, such as the metals iron and chromium. Albert Fert and Peter Grünberg won the Nobel Prize in 2007 "for the discovery of Giant Magnetoresistance." The resistivity of a metal changes with increasing magnetic field, and this effect is called the magnetoresistance. In most metals, the percentage change in resistance is only a few percent, but Fert and Grünberg found that in multilayer structures composed of alternating regions of iron and chromium, the magnetoresistance could reach 60%. This discovery has important technological implications. Magnetic storage media store information as a sequence of magnetic domains that represent ones and zeros (think of the domains as little bar magnets that point up or down). To read the information, the read head senses the variation in the orientation of the domains in the spinning disk. Magnetoresistance is used to convert changes in magnetic field above the disk into changes in resistance that can be detected electronically. The discovery of giant magnetoresistance greatly increased the sensitivity of the read head and enabled the magnetic recording industry to greatly miniaturize magnetic disks.

The giant magnetoresistance occurs because of changes in the orientation of the magnetism in adjacent layers in a multilayer structure with increasing external magnetic field. The resistance is lower if adjacent layers are aligned, and it is higher if the layers are opposed. If the adjacent layers in the read head are opposed, then the presence of a magnetic field in the disk causes the orientation of adjacent layers in the read head to change from opposed to aligned, leading to a change in the magnetoresistance.

Building on fundamental physics experiments done almost a century ago to prove the wave nature of particles like the electron, the field of surface science emerged. Materials scientists learned how crystals grow and developed techniques to grow complex structures, atomic layer by atomic layer. These developments led to smaller and faster transistors operating at the higher frequencies needed for today's communications systems, and to light emitting diodes and semiconductor laser diodes that are key components of lightwave communications (see Box 6.2). In addition, new phenomena have been discovered using these new man-made heterostructures and superlattices, and these discoveries resulted in four Nobel Prizes.

BOX 6.2 Light Emitting Diodes (LEDs) and Semiconductor Lasers

Before leaving this section on the development of epitaxial growth of materials, a brief discussion of the application of these techniques to make LEDs (light emitting diodes) and LDs (laser diodes) is in order. These light sources are essential components in optical communications and are ubiquitous in consumer electronics and barcode scanners.

Solar cells and photodiode detectors of radiation were discussed in Chapter 5. In these devices the absorption of light creates electron-hole pairs in silicon and other semiconductors. An electric current results if there is a *p–n* junction to facilitate the separation of charge before recombination occurs.

The reverse situation–the recombination of electron-hole pairs at a *p–n* junction thereby emitting light—is the basis for light emitting diodes. In turn, if the *p–n* junction grows in an optical cavity that contains the light between mirrors, then a semiconductor laser can be made.

When an electron-hole pair recombines, the energy is either emitted as light or converted into thermal energy in the form of the vibration of the atoms (phonons). These are called radiative and non-radiative decay respectively. Semiconductors such as GaAs have particularly favorable ratios of radiative vs. non-radiative decay because of subtle differences in the electronic structure. (Silicon does not have a favorable ratio, but extensive research is continuing in the hope that a way will be found using nanotechnology to improve the ratio. It would be advantageous to have computer chips that also generate and receive lightwaves to communicate with other parts of the computer or with optical fibers.)

(*cont.*)

BOX 6.2 Continued

When a p–n junction is forward biased (the p-side is made positive relative to the n-side), then a current flows and electron-hole pairs form. A large fraction of these recombine at the p–n junction emitting light. However, the light radiates in all directions.

In optical communications, the light is directed into the end of an optical fiber. Confinement of the emitted light in epitaxially grown structures is used to increase the efficiency of LEDs and to focus the light emission to some degree. (These concepts were first used in semiconductor lasers as discussed below.) The energy levels in a double heterostructure LED are similar to those in a semiconductor laser and are illustrated in Fig. 6.6. The electron-hole recombination in a thin layer of GaAs is maximized by depositing it in between regions of n-AlGaAs and p-AlGaAs. Electrons arriving from the n-AlGaAs are blocked by an energy barrier from entering the p-AlGaAs. Similarly, the holes coming from the p-AlGaAs are blocked from entering the n-AlGaAs. This increases the probability of electron-hole recombination occurring in the GaAs layer.

A bonus that results from this design is the confining or guiding of the emitted light inside the GaAs film. The index of refraction of GaAs is larger than that of AlGaAs and, as is discussed in the next chapter, this leads to total internal reflection so the emitted light is confined to the GaAs layer. A double heterostructure LED is a great improvement over a simple LED.

The real breakthrough was the invention of the laser and the development of semiconductor lasers. The first solid state laser using ruby was built by Theodore Maiman at Hughes Research Laboratory in 1960, and the first lasers based on p–n junctions were reported by four different groups within a month of each other in 1962. These included groups at General Electric (Schenectady) (Hall et al. 1962), IBM (Nathan et al. 1962), General Electric (Syracuse) (Holonyak and Bevacqua 1962), and Lincoln Laboratories (Quist et al. 1962). These lasers operated at low temperatures, and they only operated in pulse mode. To become an integral part of lightwave communications systems, semiconductor lasers had to be developed that operated continuously (CW) at room temperature.

The concept of making a sandwich of different materials so as to achieve confinement of the electron-hole pairs in the central region of the semiconductor laser or a heterostructure laser was suggested by Kroemer at Varian (Kroemer 1963) and Alferov at the Ioffe Institute (Alferov et al. 1967). By 1969, room temperature operation had been achieved (Hayashi et al. 1969 and Alferov et al. 1969). Alferov and Kroemer shared the Nobel Prize in Physics in 2000 "for developing semiconductor heterostructures used in high-speed and opto-electronics" (the other half of the prize went to Jack S. Kilby "for his part in the invention of the integrated circuit.")

The difference between a light emitting diode and a semiconductor laser diode is whether the emission of the light is spontaneous or stimulated respectively. As the word implies, spontaneous radiation occurs randomly, and the intensity of the light that is emitted is given by the number of electron-hole pairs multiplied by the recombination rate.

In stimulated emission, light passing through the laser cavity stimulates the recombination of additional electron-hole pairs. As a result of this coordinated emission, the light is coherent. To achieve stimulated emission, mirrors are put on each end of the cavity, and the light reflects back and forth in the cavity. With each pass, there is further stimulated emission so the intensity of the light grows exponentially. To achieve stimulated emission, a high concentration of electron-hole pairs needs to be confined to the active region.

In one of the directions perpendicular to the laser beam, this is done by the type of confinement shown in Fig. 6.5. A heterostructure is formed using n- and p-type $Ga_{0.7}Al_{0.3}As$ with GaAs in the middle. In the other perpendicular direction, the structure is physically made as narrow as possible. To further guide the resulting stimulated radiation as it passes through the cavity, outer layers of $Ga_{0.5}Al_{0.5}As$ are deposited. As discussed above and in the next chapter, the difference in refractive index leads to total internal reflection. So there is internal confinement of the electron-hole pairs, and there is outer confinement of the resulting light.

Laser diodes based on p–n junctions in GaAlAs/GaAs are only one example. The wavelength of the laser depends on the bandgap of the core material. By varying the composition of a semiconductor alloy, the bandgap can be tuned to build lasers with the desired wavelength. The quartz fibers used in lightwave communications have specific regions where they are most transparent, and lasers have been designed to emit light at those wavelengths. There are three windows at 850 nm, 1300 nm, and 1550 nm, and to cover this range of wavelengths, double heterostructure lasers are made using InGaAs and InGaAsP in addition to AlGaAs.

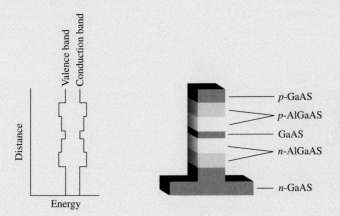

Figure 6.5 A p-AlGaAs/GaAs/n-AlGaAs laser diode. The active area for the recombination of electron-hole pairs is the GaAs in the center. The electrons and holes are confined by the energy barrier resulting from the different band gaps of GaAs and AlGaAs as shown on the left. The light resulting from the recombination of electrons and holes is confined in the vertical direction by total internal reflection at the interface between two different regions of p-AlGaAs and of n-AlGaAs because of differences in the refractive index resulting from a change in the ratio of Al to Ga in the middle of each region. A typical cross-section for the active region of the laser diode is 0.1 micrometers by 10 micrometers, that is an order of magnitude smaller than the diameter of a human hair.

7
Through the Looking Glass

Today we telephone anywhere in the world. We download music, movies, and books from the Internet without giving it a second thought. The calls and the data are carried to us, at least part of the way, as lightwaves traveling through quartz fibers. The revolution in information technology that has taken place over the past 30 years is based in part on making extremely pure quartz from sand, and then selectively doping it to make quartz fibers that guide the light for thousands of kilometers (Fig. 7.1).

An enormous amount of data is transmitted through optical fibers. Information is transmitted in the form of a string of ones and zeros or bits. A single quartz fiber has the capacity to transmit information at a rate of 1 terabit per second (1000 billion bits per second). This rate is almost a billion times higher than the capacity of a telephone line in the 1950s. This chapter is about how the telecommunication industry achieved a billion-fold increase in capacity using quartz fibers made from sand.

To try to appreciate a terabit per second of data, consider the start of the New York Marathon where just a few tens of thousands of people run over the Verrazano Narrows Bridge. Suppose that each of those runners, our hypothetical bits, could run at the speed of light. With a little crowd control, they could all pass by the entrance to the bridge in a second. This stream of people or bits is still a 100 times less than the bit rate needed to transmit a single digital telephone call. Now think of lining up all the people in the world (6.5 billion) in a ring around the Earth that passes through the entrance to the Verrazano Bridge. To match the rate of data that can pass through a single optical fiber, all the people in the world would each have to pass through the entrance to the bridge 1500 times each second.

In addition, instead of passing through the entrance to the Verrazano Bridge, the light passing through the optical fiber is confined to its core that has a diameter 8 times thinner than a human hair. Add up all the fiber optic systems that connect all the switching centers in the world, and the amount of information transmitted through quartz fibers is truly staggering.

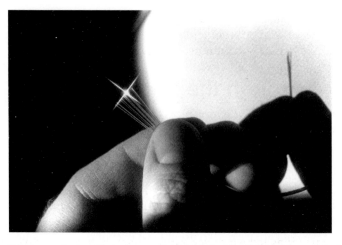

Figure 7.1 An optical cable composed of 24 hair-thin lightguides carried voice, video, and data traffic under the streets of Chicago in the world's first lightwave communications system in 1971. The optical fibers are made from sand. (Reproduced with the permission of Alcatel-Lucent USA, Inc.)

How is information transmitted in telecommunication systems and what limits the rate at which information can be transmitted? We begin by reviewing the history of the rate increase during the past fifty years and discussing the transition from electrons to photons. Next, we outline the history of confining light inside water jets to optical fibers. Finally, we discuss the production of optical fibers.

An AM radio station transmits at an assigned frequency called the carrier frequency, for example at $f_1 = 700$ kHz. To transmit information, for example the musical note A with a frequency $f_2 = 440$ Hz, the amplitude of the carrier frequency is modulated at the frequency of the note. The resulting signal contains frequencies that range from the carrier frequency plus and minus the modulation frequency and this range is referred to as the bandwidth. The human voice is composed of a range of frequencies that extends to about 4000 Hz, so a radio station or a telephone signal has a bandwidth of 8000 Hz.

The microphone in a telephone converts the sound of the human voice into a voltage that varies with time. In modern communications systems, this fluctuating voltage is in turn converted into a digital signal composed of a series of pulses representing ones and zeros. The bandwidth of a digital telephone signal is 1 MHz.

What are the limits to how fast data can be transmitted? This question was first asked in the mid-1800s with regard to the maximum speed of telegraph transmission. A telegraph signal is a form of amplitude modulation in which the modulating signal is a train of pulses. How many pulses can be sent in a given time? The individual pulses can be made shorter and spaced closer together, but there must be some limit. Above that limit, the receiver at the other end of the line cannot cleanly separate the pulses. In 1928, Harry Nyquist working at Bell Labs extended the analysis of telegraph signals to include pulses of arbitrary shape (Nyquist 1928). If the telegraph line transmits signals with frequencies up to a maximum frequency W, he proved that the maximum pulse rate is limited to twice the bandwidth 2 W.

To increase the capacity of the telecommunications network, the bandwidth has to be increased, and at the same time the frequency of the carrier wave has to increase. (The bandwidth is usually designed to be of the order of 10% of the carrier frequency.) The history of the telecommunications industry over the past half century is one of developing systems that operate at higher and higher frequencies so as to increase the bandwidth.

The mid-twentieth century telephone system transmitted information over pairs of copper wires. The capacity of a pair of copper wires increases with the frequency of the carrier wave, but there are radiation losses with increasing frequency. The practical limit for manageable radiation loss is about 1 MHz. The frequency range can be increased by changing from a pair of twisted copper wires to a coaxial cable in which a central wire is encased in a concentric cylindrical outer conductor. This extends the range of frequencies to about 100 MHz. The next step is to move to microwave frequencies, which extended the range to 100 GHz. However, microwave signals are attenuated by the air, and this limits the distance between microwave towers.

A system had to be developed that could transmit information reliably over longer distances. In the nineteenth century, J. J. Thomson and Lord Rayleigh had shown that electromagnetic waves can propagate inside hollow metallic tubes called waveguides. The development of radar during the Second World War provided expertise in the design and use of waveguides to transmit microwave signals. A heroic effort was made in the 1970s to develop waveguides that could transmit telecommunications signals over long distances at frequencies of 100 GHz. AT&T built a waveguide system from New York to Washington, which was an engineering tour de force.

Over a 20-year period after the Second World War, telecommunications companies increased the available bandwidth from thousands of Hz to tens of gigahertz, but it would have been a daunting task to implement waveguide systems across the country, to say nothing of between continents. To keep up with demand, a new way had to be found to increase frequency and bandwidth.

Fortunately, by this time work at Standard Telecommunications Laboratories in England and at Corning Glass in the US demonstrated the feasibility of transmitting information in the form of lightwaves through quartz fibers instead of using electrons to transmit information through copper wires. This increases by orders of magnitude the frequency (50 THz) and, concurrently, the bandwidth available for the reliable transmission of telecommunications signals over long distances.

Light has been used to transmit information, albeit rather slowly, for centuries. Armies used strings of fires to warn of the encroachment of enemy forces. The Boston patriots used lanterns (one if by land or two if by sea), and the navy used flashing lights in Morse code between ships.

The idea of the amplitude modulation of a beam of light as a means of transmitting voice communications was actually first demonstrated by Alexander Graham Bell a few years after he patented the telephone. In 1880, he patented the "Photophone" in which a mirrored diaphragm converted acoustic vibrations into a modulated light beam (Bell 1880). However, transmission of light through the air over long distances is too dependent on atmospheric conditions, so the photophone remained just a curiosity.

For the next century long distance communication was via copper wires and radio waves. Practical lightwave communications only became a reality in the 1980s with the confluence of three technological breakthroughs: light sources in the form of LEDs and semiconductor lasers, high speed photodiode detectors, and quartz fibers.

To be practical elements of a communication's system, optical fibers must transmit the light with minimum losses. The light is attenuated both because of the absorption of the light in the quartz and because it leaks out of the fiber as it passes through. The fiber has to be designed to guide the light with minimum loss from one end to the other. The story of the development of fiber optics is told by Jeff Hecht in his book *City of Light* and the short summary below is taken from his book (Hecht 2004).

The first published reports of the guiding of light inside jets of water were by Daniel Colladon and John Tyndall (Colladon 1842 and Tyndall 1854). Colladon's water jets were used by the Paris Opera as part of the stage set for the opera *Faust* in 1853. By confining red light inside a jet of water, the devil appeared to produce a stream of fire from a wine barrel (Hecht 2004, p. 16). A number of expositions that took place around the end of the nineteenth century featured multicolored fountains. Lights of different colors were placed under the fountain, and the different colored lights were contained within the streams of water as they rose from the fountain. As Jeff Hecht points out, it is one of those unfortunate quirks of history that Colladon's experiments became buried in the academic literature, and as a consequence Tyndall is credited with the discovery of the guiding of light (Hecht 2004, p. 27).

Light is guided along a water jet or an optical fiber by what is called total internal reflection. The guiding of light is the logical extension of a phenomenon that we often experience, namely the reflection and refraction of light. We reach into the ocean to pick up a shell resting on the sandy bottom, and we have to correct for the fact that the light changes direction when it passes from the air into the water. Our brain expects the light to continue moving in a straight line when it enters the water and tells our hand to go to what turns out to be the wrong place.

The Greek scientist, Claudius Ptolemy, was the first person to record how the angle of a light ray changes when it leaves the water. Ptolemy is best remembered for his Earth-centered model of the universe. His model predicted the motions of the heavens with reasonable accuracy with a complicated clockwork of wheeling motions around the Earth. The Copernican model with the Sun at the center of the solar system replaced the Ptolemy model and led to the well known dispute between the Roman Catholic Church that accepted the Ptolemy model and Galileo Galilei who supported the Copernican model.

Ptolemy measured the angles made by a light ray with respect to a line that is perpendicular to the surface of the water as the light passed from air into water. In AD 140 he recorded a table of the angle in air and the corresponding angle in water, as the angle in air is increased from looking almost straight down on the water (10°) to close to the surface (80°). He observed that the change in direction increased from 2° to 30° as the angle of incidence increased from 10° to 80°. As Feynman points out, this is one of the few times in Greek science that the results of an experiment were recorded (Feynman 1975, pp. 26–1).

Ptolemy didn't understand the relation between the angles in air and water, and it wasn't until 1621 that Willebrord Snell showed that the two angles are related by what is now known as Snell's Law ($n_a \sin\theta_a = n_b \sin\theta_b$). At the interface between two materials like water and air, the products of the index of refraction multiplied by the sine of the angle that the light makes with respect to a line perpendicular to the surface are equal for each component.

The refractive index is the ratio of the speed of light in a vacuum divided by the speed of light in a material. The refractive indices for air, water, and quartz are 1.0, 1.3, and 1.5 respectively. Light moves 25% and 33% slower in water and quartz respectively than it does in air. When we reach for the seashell, it is actually closer to us than we expected. The light coming from the shell bends down, as illustrated in the top left of Fig. 7.2 (light coming from the right bends down as it passes through the surface). Because the index of refraction of water is larger than that of air, the angle the ray of light makes from the perpendicular to the surface is smaller in the water than it is in the air.

The interesting thing with regard to guiding light in a water jet or optical fiber is what happens when we consider a ray of light passing from water into air. As the angle increases, there is a critical angle at which the angle in the air, which increases faster, reaches 90°. At this angle, the ray of light emerging into the air is parallel to the surface of the water. If the angle in the water is increased further then the light is reflected back into the water (total internal reflection) as shown in the top right of Fig. 7.2. For water, the critical angle is 50.3° and for quartz it is 41.8°. Total internal reflection occurs in a material if the index of refraction of the surrounding material is smaller than that of the core material.

Before moving on to discuss the guiding of light in an optical fiber, it is worth mentioning that Snell's Law follows from one of the fundamental principles in physics. This is the principle of least time developed by Pierre Fermat, a French mathematician, in 1650. The principle is that light picks the path that takes the least time in traveling from point A to point B. Feynman gives the example of a lifeguard at the beach who sees someone in trouble in the water. He knows that he can run faster than he can swim, and he realizes that he can reach the drowning man faster if he runs part way along the beach before entering the water to swim to the drowning person. Snell's law is a geometrical construction of the principle of least time for light traveling from one material to another when the speed of light is different in each of these materials. The science of geometric optics that describes the operation of lens, mirrors, and so on evolves from Fermat's principle (Feynman et al. 1975, pp. 26–4).

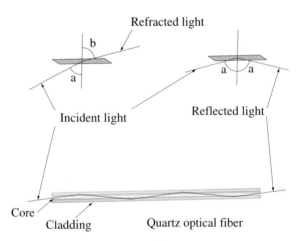

Figure 7.2 There is a critical angle for the refraction of light as it passes from water into air. Below the critical angle the light is refracted at the surface (top left) and above the critical angle it is reflected back into the water at the surface (top right). In an optical fiber (bottom), the refractive index of the core is larger than that of the cladding, and the incident light experiences total internal reflection as it passes through the fiber.

In the last half of the nineteenth century and the early part of the twentieth century, experiments with guiding light shifted from water jets to glass rods and fibers. The interest was in developing imaging devices that could transfer an image from one place to another. Doctors wanted to find ways to look inside the human body, and today we all profit from the availability of flexible instruments for examining internal organs.

An image can be transferred from one place to another by assembling a collection of parallel rods or fibers. Each fiber transmits a small piece of the image, and the resolution of the resulting image increases as the diameter of the individual fibers decreases. The trick is to preserve the orientation of each fiber with respect to the other fibers so as not to distort the image.

A number of people developed schemes for transmitting images, but the key patent entitled "Picture Transmission" is assigned to C. W. Hansell who was working at RCA Laboratories (Hansell 1927). These early fiber bundles only worked over short distances because light leaked from one fiber to another where they touched each other.

The breakthrough came with the recognition that if one clad the central fiber with a material that had a lower index of refraction, then the light is contained through total internal reflection. Although with hindsight this seems obvious, it took over 20 years before several people realized that cladding was the key element for the successful development of fiber optics. As with many important ideas, there is controversy over who had the idea first, but Hecht attributes the discovery to Brian O'Brian at the University of Rochester in 1951. Shortly after that, in 1956, the first flexible gastroscope employing clad glass fibers was developed at the University of Michigan by Basil Hirschowitz and co-workers (Hirschowitz *et al.* 1956).

Behind this success was the clever idea of his undergraduate student Lawrence E. Curtis of how to make a practical clad fiber (Curtis 1957). He put a rod of glass inside a tube of a different glass that had a lower refractive index. He heated the tube so that it collapsed onto the rod. This produced what is now called a preform. In order to draw a fiber, the preform is mounted vertically and the bottom is heated in a furnace until the glass just starts to melt. As illustrated in Fig. 7.3, the molten glass is drawn into a fiber and wrapped around a spool (in Curtis' case the spool was an old oatmeal container) as the preform is translated into the heater. As the preform necks down to form the fiber, the cross-section of the fiber scales uniformly. The resulting fiber retains a sharp cylindrical interface between the glasses of differing indices of refraction. The light passing through the core experiences total internal reflection, as illustrated in Fig. 7.2. Hirschowitz and co-workers assembled a working gastroscope from an ensemble of clad glass fibers in 1957. They then licensed American Cystoscope Makers Inc. to manufacture

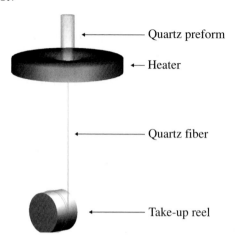

Figure 7.3 Optical fibers are drawn by heating the end of the preform, which is slowly translated through the heater. As the perform melts, the fiber is drawn and continuously taken up by the take-up reel below. The ratio of the diameter of the core to the diameter of the cladding is preserved in the drawing process.

their instrument, and their gastroscope became quite popular in the medical profession during the 1960s.

Although the technology was available by the late 1950s to make clad glass fibers, lightwave communications did not yet garner much interest from the telecommunications industry because the absorption of the light in the fiber was too large.

Since the early days of the telephone, the intensity and the attenuation of audio signals have been measured in units of decibels or dB (the Bel was chosen in honor of Alexander Graham Bell). The ear is sensitive to extremely small percentage changes in intensity independent of the absolute intensity of the sound, so the decibel is defined as the logarithm of the ratio of the intensity relative to a standard intensity. The loss of power in an optical fiber is usually quoted as dB per kilometer. If the intensity of the signal is reduced to 1% of its original level after passing through a kilometer of fiber, then it has decreased by 20 dB/km.

The first clad fibers were made out of ordinary glass that is relatively impure. The attenuation is typically 1000 dB/km (after a kilometer the fraction of the power remaining is 10^{-100}). It is one thing to make a medical instrument that is a couple of feet long, but it is quite another to transmit an optical signal many kilometers. None the less, a group at Standard Telecommunications Laboratory (STL) in England started to explore the possibilities of using glass fibers for optical communications.

A critical paper was given in 1966 by Charles Kao of STL. It laid out a general design for an optical communications system. He suggested that if the losses could be lowered to below 20 dB/km, then fiber optics would be a viable technology in the telecommunications industry (Kao and Hockham 1966). Charles Kao shared the Nobel Prize in 2009 for his early work on optical fibers.

The Corning Glass Company had extensive experience in all aspects of the use and manufacturing of glass products, and they realized that it behooved them to keep abreast of new possible applications of glass. Robert Maurer at Corning knew that silica, SiO_2, was far purer than ordinary glass. It was harder to work with because it melted at much higher temperatures (~1600 °C), but Corning had the experience to work with silica. By 1970 Corning announced the production of silica fibers with less than 20 dB/km loss (Kapron et al. 1970). The telecommunications industry finally realized that the future lay in optical communications using clad quartz fibers.

The race was on to produce purer and purer fiber with lower and lower losses. Bell Telephone Laboratories and Corning Glass in the United States, Standard Telephones and Cables and the British Post Office Laboratory in England, Philips in the Netherlands, and Fujitsu, Nippon Telegraph and Telephone Company, Nippon Sheet Glass, and Sumitomo in Japan, among others, entered the race. Within a decade the losses went from 20 dB/km to a few tenths of a dB/km, and fiber optics was on its way to becoming the dominant transmission medium in the telecommunications industry.

Just as the semiconductor industry had learned in the 1950s, and we discussed in Chapter 3, the ability to manufacture extremely pure materials is the foundation on which the industry is built. Similarly, it is the ability to selectively dope these materials, in this case to control the index of refraction of the preform during manufacture, that makes fiber optics possible.

In 1934, Corning developed a process to make silica glass that is referred to as the "soot" process (Hyde 1934), and adaptations of this process are used by many of the companies that manufacture optical fibers. As illustrated in Fig. 7.4, $SiCl_4$ and O_2 are passed over a flame, and they hydrolyze to produce SiO_2 and Cl_2. The SiO_2 collects on the rotating target rod in the form of small particles, "soot." Dopants can be added to change the index of refraction as the "soot" is deposited. F_2Cl_6 or BCl_3 are added to decrease the index of refraction of the cladding. By controlling the amount of the dopant gas and when the gas is added to the gas stream, the shape of the change in refractive index at the interface between the core and the cladding can be precisely tailored. After depositing the soot, the cylinder is heated in a furnace so that it collapses into a solid rod of silica.

Another process, which was developed by John MacChesney and Paul O'Conner at Bell Labs in 1974, is called modified chemical vapor deposition (MCVD) (MacChesney 1977). As illustrated in Fig. 7.4, the

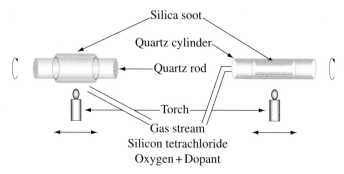

Figure 7.4 A stream of $SiCl_4 + O_2$ + dopant gas react and deposit a layer of doped SiO_2. In the Corning process (left) the layer forms on the surface of a rotating quartz rod and in the Bell Labs process (right) it forms inside a rotating quartz cylinder. In each case the rod or cylinder plus the deposited soot are heated to produce a composite cylinder, the preform, with a cladding surrounding a core in which the index of refraction of the cladding is less than that of the core.

feed gases pass through a rotating silica tube, and the chemical reaction occurs on the inside surface of the tube as an external burner passes back and forth. In the soot process, the dopant is added to the cladding to lower its index of refraction. In the MCVD process, the dopants $TiCl_4$ or $GeCl_4$ are added to the gas mixture to increase the index of refraction of the core. The deposited film is then sintered, and finally, the tube is heated to 2000 °C to collapse it to a solid preform.

Other companies have developed methods to make performs in which the method of heating and configuration of the reaction chambers vary. These include the plasma-activated chemical-vapor deposition (PCVD) developed by Phillips (Geittner *et al.* 1976) and the vapor-axial deposition method (VAD) developed by Sumitomo (Fujiwara *et al.* 1977). During the past decades, there have been a number of improvements in the different processes, but they all rely on forming ultra-pure $SiCl_4$ from semiconductor grade Si and then hydrolyzing it to make carefully doped SiO_2 preforms (Mynbaev and Scheiner 2001).

The preforms are then put into a fiber drawing apparatus, as shown schematically in Fig. 7.3. The drawing process is far more sophisticated than simply winding the resulting fiber onto an oatmeal container as Curtis did. After years of engineering, a present day fiber drawing system not only monitors the concentricity of the fiber, but it deposits a protective coating on the fiber on the way to the take-up drum. Depending on the eventual application, fiber can be drawn at rates that vary from hundreds to thousands of meters per minute.

The implementation of fiber optics into the telecommunications network went into full swing in the late 1970s and early 1980s. Hecht relates the story that the first non-experimental use of fiber optics was by the Dorset Police Department in England in 1975. Their communications system had been destroyed by lightning, and they asked Standard Telephones and Cables to build a system that would be immune to future failures of this kind. In a few weeks STC had installed a fiber optics system.

By 1977, the Bell System was sending telephone calls over a fiber link in Chicago (Fig. 7.1). The first fiber system along the northeast corridor from Boston to Washington was installed in 1983. The first transatlantic fiber cable, TAT-8, became operational in 1988. By the turn of the century close to 75 million miles of quartz fibers had been installed around the world. When the tech bubble burst in 2001, installation of fiber systems slowed, but one only has to walk down the street watching everyone either on their cell phone or text messaging or surfing the web to realize that the market for fiber optics is going to continue to grow. By the end of the first decade of the 21st century, the telecommunications companies are being hard pressed to increase the capacity of their systems to meet the real time downloading of different types of media. The information superhighway is an integral part of our lives, and that highway starts with converting sand into silicon and then into quartz fibers.

8
Sand is Everywhere

Sand is a common material that is in everything around us. Our story is about how science taught us to take this common material and provide the products on which we depend almost completely for our modern existence. These products comprise only one tenth of one percent of the sand that is used each year. In its annual report on commodity use, the US Geological Survey divides sand into construction sand and silica (high silica sand) and, in 2009, 800 million and 29 million metric tons were used respectively. (http://minerals.usgs.gov/minerals/pubs/commodity/). In contrast, about 31,000 tons of polycrystalline silicon is produced worldwide for the semiconductor and solar cell industries (Roskill 2007). We close by briefly surveying the other uses of sand and taking a tour of some of the products that contain sand in our homes and apartments.

The major uses of silica are in the glass industry, in foundry operations, and in hydraulic fracturing. The glass industry uses about 10 million tons of silica for making glass containers, windows, and fiberglass. Foundry operations use about 5 million tons to make molds for casting iron, steel, and other metals. The petroleum industry uses about 7 million tons for hydraulic fracturing. This is a process in which a slurry of water and sand is pumped into an oil well until the pressure of the slurry causes the rock to crack. In turn, the sand keeps the cracks open, thereby increasing the area from which oil can be extracted. About 4 million tons are used as fillers in paints, putty, and various building materials. In addition, sand is used in abrasives, in water filtration, on golf courses, for play sand, and numerous other applications.

Another use of silica is in the form of crystals of quartz, and these applications were discussed in Chapter 1. Quartz crystals are in electronic devices such as oscillators, filters, and timers. These devices are incorporated into communications equipment, computers, electronic games, and television sets. The USGS has a special category "quartz crystal," and as mentioned in Chapter 1, they report that 10 billion quartz plates are manufactured each year for these applications.

About half of this book discusses the use of silicon that is made from sand. The USGS breaks down the worldwide production of silicon into ferrosilicon and silicon metal. About 7,320,000 metric tons of ferrosilicon was produced in 2008, and this represents about 20% of the use of silica worldwide. At the same time, 609,000 metric tons of silicon was produced. The actual production of silicon is much larger because the number in the report excludes China and the US. (The US production is excluded to avoid disclosing company proprietary data.)

Ferrosilicon is an alloy of iron and silicon that is made by reducing sand by coke in the presence of iron. Most of the ferrosilicon (61%) goes into making cast iron, stainless steel, and other alloy steels. Most of the silicon metal (76%) is used by the chemical industry to make silanes, silicones, and other chemicals.

The list of products that have components made from sand is endless. To put them in perspective, let's look around the typical house or apartment at the non-electronic products that are derived at least in part from sand. We get up in the morning. We brush our teeth with Crest® toothpaste that contains silica. We shave using Gillette® shaving gel that contains a silicone compound. About half of all makeup, hair- and skin-care and underarm products introduced today contain silicones (http://www.dowcorning.com/content/discover). For example, some lip glosses contain a silicone acrylate copolymer (O'Lenick et al. 2007 and Martin et al. 2002), and sunblocks contain various silicones. There are bathroom and kitchen counters made of quartz composites like Zodiac® from DuPont. There are glass-ceramic cooktops like CERAN® by Schott and dishes and pots like CorningWare®. There are silicone containing cooking sprays and silicon rubber cooking utensils. When we scrub the sink, we use Comet® or when we run our dishwasher, we use Cascade®. Cheer®, used in the clothes washer, contains a silica based zeolite as do some kitty litters. For household repairs, there are silicone sealers to fill the crack between the bathtub and the tile. There is DAP® glazing compound to repair the windows. Silicones are added to paints to improve their wetting and leveling properties. Our houses have glass windows, fiberglass insulation, and sand containing asphalt roof shingles. Our kids play with silly putty that is a cross-linked polymethylsiloxane polymer. It is soft and malleable when squeezed slowly but shatters if deformed rapidly. More recently, polymers have been developed that can be used for body armor much like silly putty except that they become rigid under stress without shattering. There are the quartz envelopes used in mercury vapor lamps or tungsten halogen lights. When we go to a reception and are given a nametag, there is a silicone based release layer that allows one to remove the label from the backing so that it can be stuck on your shirt (Kennedy 2005). Silicones are used in the textile industry to improve fiber processing. Many fabric softeners incorporate silicones to promote water pick-up. They lead to breathable shirts and wrinkle-free shirts. We should also remember "That's one small step for man, one giant step for mankind" was made by Neil Armstrong on the moon wearing silicon rubber boots. This is a very incomplete tour of sand containing materials, but it shows that sand products are everywhere in our lives.

Finally, let's return to all the electronics, appliances, and toys that contain those ubiquitous silicon chips. These are made from semiconductor grade silicon, and the applications of ultra-pure silicon made up three chapters in the book. The production of polycrystalline silicon is less than a tenth of one percent of the use of silica in the United States. Twice as much sand is used for water filters in swimming pools than is used by the electronics industry, and twenty-five-times more sand is used for the greens and sand traps on golf courses.

However, the theme of this book is how that 0.1% is used in products on which we depend for our modern existence. The scientific revolution in the twentieth century enabled the technology that led to these products. Experiments to establish the wave nature of electrons and x-rays led to new tools for determining the structure of solids and their surfaces. For the first time, scientists could see inside sand based materials, and they began to understand the relation between the structure and the properties of materials. Scientists learned to control the architecture of the materials that they synthesized from sand and to grow crystals atomic layer by atomic layer. Understanding the fundamental properties of light and the emission and absorption of light led to the development of quantum mechanics. This, in turn, is at the heart of semiconductor physics that allows electrical engineers to design ever more sophisticated electronic devices based on silicon made from sand.

So when you sit on the beach and let the sand run through your fingers, I hope that you will realize how much our lives depend on those little grains of sand and how science has taught us to take this most common substance and create revolutionary technology.

Bibliography

Alferov, Z. I., Andreev, V. M., Korgokov, V. I. (1967) High voltage p-n junctions in $Ga_xAl_{1-x}As$ crystals. *Fizika I Tekhnika Polumrovodnikov* 1, 1579–82; *Soviet Physics – Semiconductors* 1, 1311–13.

Alferov, Zh. I., Andreev, V. M., Portnoi, E. L., and Trukan, M. K. (1969) AlAs-GaAs heterojunction injection lasers with a low room-temperature threshold. *Fizika I Tekhnika Poluprovodnikov* 3, 1328–32; *Soviet Physics – Semiconductors* 3, 1107–10.

Andrus, J. (1957) Fabrication of semiconductor devices. US Patent 3,122,817 – filed 8/15/1957; issued 3/3/1964.

Aruz, J., Benzel, K., Evans, J. M. (2008) *Beyond Babylon*. Yale University Press, New Haven.

Ashburn, P. (2003) *SiGe Heterojunction Bipolar Transistors*. John Wiley & Sons, New York.

Baerlocher, C., Meier, W. M., and Olson, D. H., eds. (2001) *Atlas of Zeolite Framework Types*. Elsevier, Amsterdam.

Baker, J. (2006) Look into the Seeds of Time. *Science* 314, 1707.

Barbo, L., Beaudouin, D. and Laguës, M. (2004) *L'Expérience Retrouvée*. Belin, Paris.

Bardeen, J. and Brattain, W. (1948) Three-electrode circuit element utilizing semiconductor materials. US Patent 2,524,035 – filed 6/17/1948; issued 10/3/1950.

Barlow, W. (1894) Ueber die geometrischen Eigenschaften homogener starrer Structuren und ihre Anwendung auf Krystalle. *Zeitschrift für Kristalographie* 23, 1–63.

Barlow, W. (1898) A mechanical cause of homogeneity of structure and symmetry geometrically investigated; with special application to crystals and to chemical composition. *Proc. Phys. Soc. London* 16, 54.

Bell, A. G. (1880) Apparatus for signaling and communicating called Photophone. US Patent 235,199 – filed 8/28/1880; issued 12/7/1880.

Bragg, W. L. (1913) The structure of some crystals as indicated by their diffraction of x-rays. *Proc. Roy. Soc. London* A89, 248–77.

Bragg, W. H. and Bragg, W. L. (1913) The structure of diamond. *Proc. Roy. Soc. London* A89, 277–91.

Bragg, W. and Gibbs, R. E. (1926) The structure of α and β quartz. *Proc. Royal Soc. London* A109, 405–27.

Bravais, M. A. (1850) Mémoire sur les systèmes formés par des points distributes regulièrement sur un plan ou dans l'espace. *J. école polytech.* 19, 1–128 and (1949) ACA Monograph 1, American Crystallographic Association.

Brian, D. (2005) *The Curies; A Biography of the Most Controversial Family in Science*. John Wiley & Sons, New York.

Brinker, C. J. (1996) Porous inorganic materials. *Current Opinion in Solid State & Materials Science* 1, 798–805.

Burnett, D. S. (2006) NASA Returns Rocks from a Comet. *Science* 314, 1709–10.

Cady, W. G. (1920) Piezo Electric Resonator. US Patent 1,450,246 – filed 1/28/1920; issued 4/23/1923.

Chapin, D. M., Fuller, C. S. and Pearson, G. L. (1954) Solar energy converting apparatus. US patent 2,780,765 – filed 3/5/1954; issued 2/5/1957.

Cho, A. Y. (1971) Film deposition by molecular-beam techniques. *J. Vac. Sci. Technol.* 8, S31–S38.

Colladon, D. (1842) On the reflections of a ray of light inside a parabolic liquid stream. *Comptes Rendus*, 15, 800–2.

Crabtree, G. W. and Lewis, N. S. (2007) Solar energy conversion. *Physics Today*, March 2007, pp. 37–40.

Cressler, J. D. and Niu, G. (2003) *Silicon Germanium Heterojunction Bipolar Transistors*. Artech House, Boston.

Cropper, W. H. (2001) *Famous Physicists*. Oxford University Press, Oxford, p. 43.

Curie, P. and Curie, J. (1880) Development by pressure of polar electricity in hemihedral crystals with inclined faces. *C. R. Acad. Sci. Paris* 91, 294 and (1881) 93, 1137.

Curtis, L. E. (1957) Glass fiber optical devices. US Patent 3,589,793 – filed 5/6/1957; issued 6/29/1971.

Davisson, C. J. and Germer, L. H. (1927) Diffraction of electrons by a crystal of nickel. *Phys. Rev.* 30, 705–40.

Derick, L. and Frosch, C. J. (1956) Oxidation of Semiconductive Surfaces for Controlled Diffusion. US Patent 2,802,760 – filed 12/2/1956; issued 8/13/1957.

Deshpande, R., Smith, D. M., and Brinker, C. J. (1993) Preparation of high porosity Xerogels by chemical surface modification. US Patent 5,565,142 – filed 4/28/1993; issued 10/15/1996.

Dingle, R., Wiegmann, W., and Henry, C. H. (1974) Quantum states of confined carriers in very thin $Al_xGa_{1-x}As$-$GaAs$-$Al_xGa_{1-x}As$ heterostructures. *Phys. Rev. Lett.* 33, 827–30.

Dingle, R., Stormer, H. L., Gossard, A. C., and Wiegmann, W. (1978) Electron mobilities in modulation-doped semiconductor heterojunction superlattices. *J. Appl. Phys.* 33, 665–7.

Errol, E. P., Ernisse, P., Ward, R. W., and Wiggins, R. B. (1988) Survey of quartz bulk resonator sensor technologies. *IEEE Transactions on Ultrasonics, Ferroelectrics, and Frequency Control* 35, 323–30.

Esaki, L. and Tsu, R. (1970) Superlattice and negative differential conductivity in semiconductors. *J. Res. Develop. IBM* 14, 61–5.

Everett, D. H. (1972) Definitions, terminology and symbols in colloid and surface chemistry; part I. In *IUPAC Manual of Symbols and Terminology for Physiochemical Quantities and Units*; Appendix II. Butterworths, London 31, 638–78.

Fedorov, E. (1895) Theorie der Krystallstructur. *Zeitschrift für Kristalographie* 24, 209–52 and (1896) 25, 113–224.

Feynman, R. P., Leighton, R. B., and Sands, M. (1975) *The Feynman Lectures on Physics*, vol. I, chapters 44 and 45, Addison-Wesley, Reading.

Friedrich, W., Knipping, P., and Laue, M. (1912) Interferenz-Erscheinungen bei röntgenstrahlen. *Annalen der Physik* 346, 971–988 and http://onlinelibrary.wiley.com/doi/10.1002/andp.19133461004/abstract [accessed 10 August 2011].

Fujiwara, K., Tanaka, G., and Kurosaki, S. (1977) Method of manufacturing glass for optical waveguide. US Patent 4,135,901 – filed 6/13/1977; issued 1/23/1979.

Geittner, P., Küppers, D., and Lydtin, H. (1976) Low-loss optical fibers prepared by plasma activated chemical vapor deposition (CVD). *Applied Physics Letters* 28 (11), 645–6.

Gibbs, J. W. (1876) On the equilibrium of heterogeneous substances. *Transactions of the Connecticut Academy of Science* vol. III, 108–248 and (1978) 343–524.

Gibbs, J. W. (1902) *Elementary Principles of Statistical Mechanics Developed with Especial Reference to the Rational Foundation of Thermodynamics*. Charles Scribner's Sons, New York.

Gibbs, J. W. (1961) *The Scientific Papers of J. Willard Gibbs*. Dover Publications, New York.

Gibbs, R. E. (1927) Structure of α Quartz, *Proc. Roy. Soc. London* A110, 443–55.

Green, M. (2002) *Power to the People: Sunlight to Electricity using Solar Cells*. University of New South Wales Press, Sidney.

Greenberg, Gary (2008) *A Grain of Sand*. Voyageur Press, Minneapolis.

Heising, R. A. (1946) *Quartz Crystals for Electrical Circuits; Their Design and Manufacture*. Van Nostrand Company, New York.

Hackmann, W. D. (1986) Sonar research and naval warfare 1914–1954: A case study of a twentieth-century establishment science. *Historical Studies in the Physical and Biological Sciences* 16, 83–110.

Hall, R. N., Fenner, G. E., Kingsley, J. D., Soltys, T. J., and Carlson, R. O. (1962) Coherent light emission from GaAs junctions. *Phys. Rev. Letters* 9, 366–8.

Hansell, C. W. (1927) Picture transmission. US Patent 1,751,584 – filed 8/13/1927; issued 3/25/1930.

Haüy, R. J. (1782) Mémoire sur la structure des spaths calçares. *Journal de Physic* 19, 366.

Haüy, R. J. (1784) *Essai d'une théorie sur la structure des crystaux, appliquée à plusieurs genres de substances crystallisées*. Chez Gogué & Née de la Rochelle, Paris. Available at: www.books.google.com.

Hayashi, I., Panish, M., and Foy, P. W. (1969) A low-threshold room-temperature injection laser. *IEEE Journal of Quantum Electronics* 5, 211–13.

Heaney, P. J. and Post, J. E. (1992) Widespread distribution of a novel silica polymorph in microcrystalline quartz varieties. *Science* 255, 441–3.

Heaney, P. J. (1994) Structure and chemistry of the low-pressure silica polymorphs. In Heaney, P. J., Prewitt, C. T., and Gibbs, G. V., eds. *Silica Physical Behavior, Geochemistry and Materials Applications, Reviews in Mineralogy* 29, 1–40.

Hecht, J. (2004) *City of Light*. Oxford University Press, Oxford.

Higgins, J. B. (1994) Silica zeolites and clathrasils. In Heaney, P. J., Prewitt, C. T. and Gibbs, G. V., eds. *Silica Physical Behavior, Geochemistry and Materials Applications, Reviews in Mineralogy* 29, 507–39.

Hirschowitz, B. I., Peters, C. W., and Curtis, L. E. (1956) Flexible light transmitting tube. US Patent 3,010,357 – filed 12/28/1956; issued 11/28/1961.

Hoddeson, L. and Daitch, V. (2002) *True Genius: The Life and Science of John Bardeen*. Joseph Henry Press, Washington, D.C.

Hoerni, J. A. (1959) Method of manufacturing semiconductor device. US Patent 3,025,589 – filed 5/1/1959; issued 3/20/1962.

Holonyak, N. Jr. and Bevacqua, S. F. (1962) Coherent (visible) light emission from $Ga(As_{1-x}P_x)$ junctions. *Appl. Phys. Letters* 1, 82–3.

Hunter, G. K. (2004) *Light Is a Messenger*. Oxford University Press, Oxford.

Hyde, J. F. (1934) Method of making a transparent article of silica. US Patent 2,272,342 – filed 8/27/1934; issued 2/10/1942.

Jenkin, J. (2008) *William and Lawrence Bragg, Father and Son*. Oxford University Press, Oxford.

Kahng, D. (1960) Electric field controlled semiconductor device. US Patent 3,102,230 – filed 5/31/1960; issued 8/27/1963.

Kao, K. C. and Hockham, G. A. (1966) Dielectric-fibre surface waveguides for optical frequencies. *Proc. of the IEEE* 113(7), 1151–58.

Kapron, F. P., Keck, D. B., and Maurer, R. D. (1970) Radiation losses in glass optical waveguides. *Applied Physics Letters* 17, 423–5.

Kennedy, P. R. (2005) Labelling assembly. US Patent 7,211,163 – filed 9/6/2005; issued 5/1/2007 and O'Lenick and O'Lenick (2007).

Kilby, J. S. (1959) Miniaturized electronic circuits. US Patent 3,138,743 – filed 2/6/1959; issued 6/23/1964.

Kroemer, H. (1963) A proposed class of heterojunction injection lasers. *Proc. of the IEEE* 51, 1782–3.

Langevin, P. (1920) Piezoelectric Signaling Apparatus. US Patent 2,248,870 – filed 6/21/1920; issued 7/8/1941.

Lenardic, D. (2011) www.sunenergy.eu/en/top50pv.php [accessed 10 August 2011].

Lightman, L. (2005) *The Discoveries*, pp. 84–110. Pantheon Books, New York.

Lilienfeld, J. E. (1926) Method and apparatus for controlling electric currents. US Patent 1,745,175 – filed 10/8/1926; issued 1/18/1930.

MacChesney, J. B. (1977) Optical fiber fabrication and resulting product. US Patent 4,217,027 – filed 8/29/1977; issued 8/12/1980.

Macfarlane, A. and Martin, G. (2002) *Glass – A World History*. Univ. of Chicago Press, Chicago.

Marrison, W. A. (1930) The Crystal Clock. *Proc. National Academy of Science.* 16, 496–507, http://www.pnas.org/content/16/7.toc [accessed 8 August 2011].

Martin, J.-C. (1976) Quartz Thermometer. US Patent 4,039,969 – filed 2/2/1976; issued 8/2/1977 and quartz thermometer manufactured by: HP/Agilent model 2804A.

Martin, S. R., Sandstrom, G. A., Rothouse, J. N., Anderson, G. T., Letton, A., Lin, H.-T., and Zheng, T. (2002) Aqueous cosmetic coloring and gloss compositions having film formers. US Patent 7,323,162 – filed 12/27/2002; issued 1/29/2008.

Mason, H. J. (1978) *Flint the Versatile Stone*. Providence Press, Ely.

McWhan, D. B. (1985) Structure of chemically modulated films. In Chang, L. L. and Giessen, B. C., eds. *Synthetic Modulated Structures*, Academic Press, New York, pp. 43–74.

Millman, S., ed. (1983) *A History of Engineering and Science in the Bell System: Physical Sciences (1925–1980)*, p. 417. AT&T Bell Laboratories, Murray Hill, NJ.

Milton, R. M. (1953) Molecular sieve adsorbants. US Patent 2,882,243 – filed 12/24/1953; issued 4/14/1959.

Mohta, N. and Thompson, S. E. (2005) Mobility enhancement: circuits and devices, *IEEE* 21, #5, 18–23.

Mynbaev, D. K. and Scheiner, L. L. (2001) *Fiber-Optic Communications Technology*. Prentice Hall, Upper Saddle River, pp. 210–18.

Nathan, M. I., Dumke, W. P., Burns, G., Dil, F. H. Jr., and Lasher, G. (1962) Stimulated emission of radiation from GaAs p-n junctions. *Appl. Phys. Letters* 1, 62–4.

Newsam, J. M. (1986) The zeolite cage structure. *Science* 231, 1093–8.

Nicolson, A. M. (1918) Piezophony. US Patent 1,495,429 – filed 4/10/1918; issued 5/27/1924.

Nobel, A. (1868) Improved Explosive Compound. US Patent 78,317 issued 5/26/1868.

Nyquist, H. (1928) Certain topics in telegraph transmission theory. *Trans. Amer. Inst. Elec. Eng.* 41, 617–44.

Ohl, R. S. (1941) Light-Sensitive Electric Device Including Silicon. US Patent 2,443,542 – filed 5/27/1941; issued 6/15/1948.

Ohl, R. S. (1950) Semiconductor translating device. US Patent 2,750,541 – filed 1/31/1950; issued 6/12/1950.

O'Lenick, A. J. Jr. and O'Lenick, K. A. (2007) Silicon polymers in skin care. *Materials Research Society Bulletin* 32, 801.

Orton, J. (2004) *The Story of Semiconductors*. Oxford University Press, Oxford.

Patrick, W. A. (1918) Silica gel and process for making same. US Patent 1,297,724 – filed 12/7/1918; issued 3/18/1919.

Patton, G. L., Iyers, S. S., Delage, S. L., Tiwari, S., Stork, J. M. C. (1988) Silicon-germanium-base heterojunction bipolar transistor by molecular beam epitaxy. *Dev. Lett. IEEE* 9, 165–7.

Pauling, L. (1960) *The Nature of the Chemical Bond*. Cornell University Press, Ithaca.

Perry, C. C. (2003) Silicification: the processes by which organisms capture and mineralize silica. In Dove, P. M., DeYoreo, J. J., and Weiner, S., eds. *Biomineralization – Reviews in Mineralogy and Geochemistry* 54, 291–327.

Pfann, W. G. (1951) Process for controlling solute segregation by zone melting. US Patent 2,739,088 – filed 11/16/1951; issued 3/20/1956.

Pfann, W. G. (1966) *Zone Melting*. John Wiley & Sons, New York.

Post, J. E. (1997) *The National Gem Collection*. Harry N. Abrams Inc., New York.

Quist, T. M., Rediker, R. H., Keyes, R. J., Krag, W. E., La, B., McWhorter, A. L., and Zeiger, J. J. (1962) Semiconductor maser of GaAs. *Appl. Phys. Letters* 1, 91–2.

Reeves, R. (2008) *A Force of Nature – The Frontier Genius of Ernest Rutherford*. W. W. Norton & Company, New York.

Reid, T. R. (1958) *The Chip*. Simon & Schuster, New York.

Riordan, M. and Hoddeson, L. (1997) *Crystal Fire: The Invention of the Transistor and the Birth of the Information Age*. W. W. Norton & Company, New York.

Roskill (2007) Forecast world production of multi-crystalline silicon by producer—2005–2011. *Roskill's Letter from Japan*, no. 367, March, p. 14: http://www.roskill.com/newsletters/roskills-letter-from-Japan [accessed 10 August 2011].

Rossman, G. R. (1994) Colored varieties of the silica minerals. In Heaney, P. J., Prewitt, C. T., and Gibbs, G. V., eds. *Silica: Physical Behavior, Geochemistry, and Materials Applications, Reviews in Mineralogy* Vol. 29, ch. 13.

Rubin, L. and Poate, J. (2003) Ion implantation in silicon technology. *The Industrial Physicist* June/July, 12–15 American Institute of Physics, Melville.

Sanders, J. V. (1964) Colour of precious opal. *Nature* 204, 1151–3.

Schönflies, A. (1891) *Krystallsysteme und Krystallstructur*. B. G. Teubner, Leipzig.

Shockley, W. (1948) Circuit Element Utilizing Semiconductive Material. US Patent 2,569,347 – filed 6/26/1948; issued 9/25/1951.

Shockley, W. (1954) Forming semiconductive devices by ionic bombardment. US Patent 2,787,564 – filed 10/28/1954; issued 4/2/1957.

Shockley, W. and Queisser, H. J. (1961) Detailed balance limit of efficiency of p-n junction solar cells. *J. Applied Physics* 32, 510–19.

Shurkin, J. N. (2006) *Broken Genius – The Rise and Fall of William Shockley, Creator of the Electronic Age*. Macmillan, New York.

Siever, R. (1988) *Sand*. Scientific American Library, New York.

Slaoui, A. and Collins, R. T. (2007) Advanced inorganic materials for photovoltaics. *Materials Research Society Bulletin* 32, 211–14.

Sobel, D. (1995). *Longitude*. Walker and Co., New York.

Stookey, S. D. (1956) Method of making ceramics and product thereof. US Patent 2,920,971 – filed 4/4/1956; issued 1/12/1960.

Swanson, R. M. (2006) A vision for crystalline silicon photovoltaics. *Progress in Photovoltaics: Research and Applications* 14, 443–53.

Swanson, R. M. (2009) Photovoltaics power up. *Science* 324, 891–2.

Temkin, H., Bean, J. C., Antreasyan, A., and Leibenguth, R. (1988) Ge(x)Si(1-x) strained-layer heterostructure bipolar transistors. *App. Phy. Lett.* 52, 1089–93.

Theuerer, H. C. (1952) Method for processing semiconductor materials. US Patent 3,060,123 – filed 10/17/1952; issued 10/23/1962.

Tyndall, J. (1854) On some phenomena connected with the motion of liquids. *Proceedings of the Royal Institution of Great Britain* 1, 446–8.

US Geological Survey, Mineral Commodity Summaries (2008) http://minerals.usgs.gov/minerals/pubs/mcs/2011/mcs2011.pdf [accessed 10 August 2011].

van der Merwe, J. H. (1962) Crystal interfaces Part II. Finite overgrowths. *J. Appl. Phys.* 34, 123–8.

Walker, A. C. and Buehler, E. (1950) Growing large quartz crystals. *Industrial Engineering Chemistry* 42, 1369–75 and Buehler, E. (1952) Method of Growing Quartz Crystals. US Patent 2,785,058 – filed 4/28/1952; issued 3/12/1957.

Wanlass, F. W. (1963) Low Stand-By Power Complementary Field Effect Circuitry. US Patent 3,356,858 – filed 6/18/1963; issued 12/5/1967.

Ward, M. D. and Buttry, D. A. (1990) In situ interfacial mass detection with piezoelectric transducers. *Science* 249, 1000–7.

Welland, M. (2009) *Sand: A journey through science and the imagination*. Oxford University Press, Oxford.

Wilkes, J. G. (1996) Silicon processing. In Cahn, R. W., Haasen, P., and Kramer E. J., eds. *Materials Science and Technology* 16, 1–62. VCH Publishers Inc., New York.

Wilson, E. B. (1901) *Vector Analysis, A Text-book for the use of Students of Mathematics and Physics, Founded upon the Lectures of J. Willard Gibbs*. Charles Scribner's Sons, New York.

Xu, D.-X., Shen, G.-D., Willander, M., Ni, W.-X., and Hansson, G. V. (1988) n-Si/p-Si(1-x)Ge(x)/n-Si double-heterojunction bipolar transistors. *Appl. Phy. Lett.* 52, 2239–41.

Yool, A. and Tyrrell, T. (2003) Role of diatoms in regulating the ocean's silicon cycle. *Global Biogeochem. Cycles* 17(4), 1103.

Zachariasen, W. H. and Ellinger, F. H. (1963) The crystal structure of alpha plutonium metal. *Acta. Cryst.* 16, 777–83.

Zachariasen, W. H. (1932) Atomic arrangement in glass. *J. Am. Chem. Soc.* 54, 3841–51.

Index

absorption of alpha particles 24, 86
absorption of light 66, 67
absorption of x-rays 24, 86
accelerometer 14
aerogel 33, 42
agate 65
AlAs 107–111
Alcatel-Lucent USA 33, 77, 88, 118
Alferov, Z. I. 114
amethyst 65–67
Amontons, G. 61
Andrus, J. 81
Arthur, J. 107
AT&T 72, 73
Avogadro, A. 61

bandwidth 118–120
Barbo, L. 6, 7
bar code 27, 29, 30
Bardeen, J. 74–76
barium 8
Barkla, C. 25
Barlow, W. 19, 21, 22
Barrer, R. M. 40
BCl_3 125
Becquerel, Alexandre 90
Becquerel, Henri 5, 6
Becquerel rays 5, 6
Bell, A. G. 120
Bell Telephone Laboratories 11, 12, 47, 49, 69, 72–79, 81, 82, 87, 88, 110–112, 119, 125
biomineralization 44
bismuth 8
black body radiation 91–93
bloodstone 65
Board of Invention and Research 10, 26
Bohr, N. 85
Boltzmann, L. 61
Boyle, R. W. 10
Boyle, Robert 60, 61

Boyle, Williard 100
Bragg, W. H. 23–27
Bragg, W. L. 15, 23–27
Bragg's Law 28–29, 44
Brattain, W. 73–75
Bravais lattice 19
Brinker, C. J. 42
Brookhaven National Laboratory 30, 99, 100
Buehler, E. 11

Cady, Walter G. 10
carnelian 65
Carnot, Sadi 54, 55, 62
Cavendish Laboratory 27
CCD detector 98–100
CdS/CdTe 96
CdS/CdS/InP 96
ceramics 36
Ceran® 37
Chapin, Daryl 94
characteristic x-rays 25
Charles, J. A. C. 61
Chilowsky, Constantin 9
Cho, A. Y. 107
citrine 65–67
Clausius–Clapeyron equation 55, 64
coesite 15
Colladon, D. 120
Comet® 130
Comet "Wild 2" 42
complementary metal oxide semiconductor (CMOS) 72, 78, 79
conduction band 68, 70–72
Cooper, L. 76
copper sulfate 23
Corelle® 37
Corning Glass Company 38, 125
CorningWare® 37, 130
Crest® 130
Crick, F. 27
cristobalite 15

Cronstedt, A. 38
Curie, Jacques 3
Curie, Marie 5–8
Curie, Pierre 3–8
Curie temperature 5
Curtis, L. E. 123
Czochralski crystal growth 51, 52

Darwin, C. G. 25
Davisson, C. 101, 102
deBroglie, L. 102
Derick, L. 79, 80
diatoms 44–46
diatomaceous earth 46
Dingle, R. 111
diopside 31
directional bonds 34
dislocations 51
Dorset Police Department 126, 127
dynamite 46

Einstein, A. 93
electrical conductivity 68, 70–72
electrometer 4
electron-hole pair 94, 95, 113, 114
energy band 68, 70–72
energy gap 68, 70–72, 96, 109–111, 114, 115
entropy 64
epitaxy 106–109
Esaki, L. 109

Fairchild Semiconductors 76–78
Faujasite 39, 41
F_2Cl_6 125
Fedorov, E. 19
ferrosilicon 130
Fert, A. 112
Feynman, R. P. 64, 122
field effect transistor (FET) 77, 78, 104
Fisk, J. 74
flint 33, 36
float zone crystal growth 51
Fourier analysis 31
Friedel, Charles 3
Friedrich, W. 23
Fritts, Charles 90
Frosch, C. 79, 80
frustule 45
Fuller, C. 94

GaAs 107–111
gastroscope 123, 124
Gay-Lussac, J. L. 61
$GeCl_4$ 126

Geiger, H. 84
General Electric 114
germanium 48, 49, 77
giant magnetoresistance 112, 113
Gibbs, J. Willard 56, 57
Gibbs, R. E. 15
Gillette® 130
glass 35
glass ceramics 37, 38
Gossard, A. C. 110, 112
Green, M. 95
Grünberg, P. 112

Haeckel, Ernst 45
Hansell, C. W. 123
Haüy, Rene 18, 19
Hecht, J. 120
Heisenberg, W. 85
heteroepitaxi 108
heterostructure bipolar transistor (HBT) 105
Hirschowitz, B. 123
Hoerni, J. R. 78
homoepitaxi 108
hydraulic fracturing 129
hydrophone 8
hydrothermal synthesis 11

IBM 114
index of refraction 115, 121, 122
InGaAs 115
InGaAsP 115
integrated circuit 78, 79
INTEL 76, 80, 104
Ioffe Institute 114
ion implantation 81, 82, 85

jasper 65
junction transistor 76–78, 104, 105

Kahng, D. 77
Kao, C. 124, 125
Kelly, M. 74
Kendrew, J. C. 27
Kilby, J. 78, 79
Knipping, P. 23
Kroemer, H. 114

Langevin, P. 8–10
Laughlin, R. B. 112
LEED 101, 103
Lewis, G. N. 17, 34
light emitting diode (LED) 113, 114, 120
Lilienfeld, J. E. 77

Lincoln Laboratory 114
Lippman, G. 4
liquid phase epitaxy 106
lithography 81, 82

MacChesney, J. B. 125
Maiman, T. 114
Marrison, Walter 12
Marsden, E. 84
Mars Rover 42
Maxwell, J. C. 61
metal-organic vapor phase epitaxy (MOVPE) 106
metal oxide semiconductor field effect transistor (MOSFET) 77, 82
micro-electromechanical systems (MEMS) 14
Milton, R. 40
misfit dislocations 108
mobility 104, 105, 111
modified chemical vapor deposition (MCVD) 125
molecular beam epitaxy (MBE) 106, 107
Moore, G. 76
Moore's Law 80, 103
Mosley, G. J. 25
multilayer structures 109–112

NaCl 22
National Synchrotron Light Source 30
Nicolson, Alexander McLean 10
Nissen, H.-U. 44
Nobel, A. 46
Nobel Prize 3, 8, 23, 26, 27, 72, 76, 84, 93, 103, 111, 112, 114, 125
Noyce, R. 78
n-type 68–72
Nyquist, H. 119

O'Brian, B. 123
O'Conner, P. B. 125
Ohl, Russell 69, 70, 75, 94
onyx 65
opal 33, 43, 44
optical fibers 33, 117, 118, 120, 123–126
Orton, J. 79

Patrick, W. 41
Pauling, L. 17, 34
Pearson, G. 94
Perutz, M. 27
Petroff, P. 110
Pfann, W. G. 47–49
phase diagram 59
phase rule 56

phosphorescence 5
photodiode 97, 120
photoelectric effect 93
photon 66, 93
Photophone 120
photovoltaic 88, 89
piezoelectricity 2–5, 15–18
p–i–n diode 98
pitchblende 7, 8
planar technology 78
Planck, M. 92
Planck's Law 91–92
plasma-activated chemical-vapor deposition 126
plutonium metal 32
p–n junction 69–72, 87, 94–96
point groups 18
polonium 8
preform 123–126
Ptolemy, C. 121
p-type 68–72
pyroelectricity 3

quantum confinement 111, 112, 114, 115
quantum of energy 90–93
quartz 1, 2
quartz accelerometers 14
quartz clocks 12, 13
quartz crystal microbalance (QCM) 13
quartz optical fiber 117, 120, 124, 125
quartz resonator 10
quartzite 47
quartz, structure 15–17
quasicrystal 19

radiation detectors 97–100
radiative decay 113
radioactivity 7, 8
radiolarians 44
radium 8
Rayleigh Jeans law 91, 92
Raoult, F. M. 58
recombination 71, 95, 97, 113
reflection 121, 122
reflection high energy electron diffraction (RHEED) 101, 103, 110
refraction 121, 122
Relativistic Heavy Ion Collider (RHIC) 99, 100
rock crystal 66, 67
Röntgen, W. 22
Rutherford, Ernest 25, 83, 84

Sameshima, J. 40
sand 1, 21, 32, 34, 47, 50, 60, 87, 117, 129–131
Sanders, J. V. 43
sardonyx 65
Scaff, Jack 70
Schönflies, A. 19
Schreiffer, R. 76
Schrödinger, E. 85
scolecite 33
semiconductor laser 114, 115, 120
Shechtman, D. 19
Shockley, W. 72–76, 82
Shockley Semiconductor 76
Shurkin, Joel 74
$SiCl_4$ 126
Siemens process 50
SiGe alloy 104, 105
silanes 130
silicic acid 36, 45
silica 1, 2
silica gel 33, 41, 42
silica tetrahedrons 34
silicon, amorphous 96
silicon cycle 45
silicones 130
silicon oxide (SiO_2) 1, 2, 79
silicon rectifiers 68–70
silicon solar cells 88, 90, 94, 95
silicon, structure 28–31
silly putty 130
Smith, G. E. 100
Smith, Willoughby 90
Snell, W. 121, 122
Sobel, D. 12
sodium silicate 41
solar spectrum 90, 92
soot process 125
space groups 15, 19
spontaneous emission 114
Square Paul Langevin 10
Standard Telecommunications Laboratories 124
Standard Telephones and Cables 127
STAR detector 99, 100
Stefan–Boltzmann Law 91
Stephens, Peter 30
stimulated emission 114, 115
stishovite 15
Stookey, S. 38
Stormer, H. L. 112
superconductivity 76
supercritical carbon dioxide 42
surface states 75, 79
symmetry 2, 15, 18–20

TELSTAR 87, 88
Texas Instruments 78
Theuerer, Henry 49
thermodynamics 54, 56, 60, 61–64
thermodynamics, first law 54, 63, 64
thermodynamics, second law 54, 55, 61–64
Thomson, G. 102, 103
Thomson, J. J. 24, 83, 102
$TiCl_4$ 126
tortional balance 4
total internal reflection 121, 122
transistor 75–77
trichlorosilane 50, 51
tridymite 15, 35
trimethylsilane 42
triple junction solar cell 96
Tsu, R. 109
Tsui, D. C. 112
two-dimensional electron gas (2-DEG) 111, 112
Tyndall, J. 120

US Geological Survey 11, 129, 130

valence band 68, 70–72
van der Merwe, J. H. 108
Vanguard I 87
Vapor-axial deposition 126
vapor phase epitaxy (VPE) 106
vapor pressure 53, 55, 58
Varian 114
Visions® 37
von Klitzing, K. 112
von Laue, M. 23

Walker, A. C. 11
water jets 120
Watson, J. 27
Western Electric Company 10, 72
Wien Displacement Law 91
Winlass, F. 77

x-rays 22, 23
x-ray diffraction 27–32
x-ray powder diffraction 29, 30

Zachariasen, W. H. 32, 35
zeolite 38–41
zinc sulfide 23
Zodiac® 130
zone refining 48–50